COLD TOLERANT MICROBES IN SPOILAGE AND THE ENVIRONMENT

THE SOCIETY FOR APPLIED BACTERIOLOGY
TECHNICAL SERIES NO. 13

COLD TOLERANT MICROBES IN SPOILAGE AND THE ENVIRONMENT

Edited by

A. D. RUSSELL

Welsh School of Pharmacy, University of Wales Institute of Science and Technology, Cardiff, UK

AND

R. FULLER

National Institute for Research in Dairying, Shinfield, Reading, UK

1979

ACADEMIC PRESS
LONDON · NEW YORK · SAN FRANCISCO
A Subsidiary of Harcourt Brace Jovanovich, Publishers

ACADEMIC PRESS INC. (LONDON) LTD
24/28 OVAL ROAD
LONDON NW1

U.S. Edition Published by
ACADEMIC PRESS INC.
111 FIFTH AVENUE
NEW YORK, NEW YORK 10003

British Library Cataloguing in Publication Data

Society for Applied Bacteriology. *Autumn Demonstration Meeting, Brunel University, 1977*
Cold tolerant microbes in spoilage and the environment.—(Society for Applied Bacteriology. Technical series; no. 13).
1. Food—Microbiology—Congresses
2. Food contamination—Congresses
3. Psychotropic organisms—Congresses
4. Microbial ecology—Congresses
I. Title II. Russell, A D III. Fuller, Roy, *b. 1933* IV. Series
576′.163 QR115 79–50315

ISBN 0–12–603750–7

Printed in Great Britain by
Latimer Trend & Company Ltd, Plymouth

Contributors

B. W. ADAMS, *A.R.C. Food Research Institute, Colney Lane, Norwich NR4 7UA, UK*

J. H. BAKER, *River Laboratory, Freshwater Biological Association, East Stoke, Wareham, Dorset BH20 6BB, UK*

ELLA M. BARNES, *A.R.C. Food Research Institute, Colney Lane, Norwich NR4 7UA, UK*

M. BHAKOO, *Department of Biological Sciences, The University, Dundee DD1 4HN, UK*

CHRISTINA M. COUSINS, *National Institute for Research in Dairying, University of Reading, Shinfield, Reading RG2 9AT, UK*

R. H. DAINTY, *A.R.C. Meat Research Institute, Langford, Bristol BS18 7DY, UK*

F. L. DAVIS, *National Institute for Research in Dairying, University of Reading, Shinfield, Reading RG2 9AT, UK*

CHARMAIGNE D. HARDING, *A.R.C. Meat Research Institute, Langford, Bristol BS18 7DY, UK*

R. A. HERBERT, *Department of Biological Sciences, The University, Dundee DD1 4HN, UK*

C. S. IMPEY, *A.R.C. Food Research Institute, Colney Lane, Norwich NR4 7UA, UK*

B. A. LAW, *National Institute for Research in Dairying, University of Reading, Shinfield, Reading RG2 9AT, UK*

J. V. LEE, *Public Health Laboratory, Preston Hall Hospital, Maidstone, Kent ME20 7NH, UK*

J. W. LEFTLEY, *Dunstaffnage Marine Research Laboratory, P.O. Box 3, Oban, Argyll PA 34 4AD, UK*

VALERIE M. MARSHALL, *National Institute for Research in Dairying, University of Reading, Shinfield, Reading RG2 9AT, UK*

G. C. MEAD, *A.R.C. Food Research Institute, Colney Lane, Norwich NR4 7UA, UK*

SILVIA MICHANIE,* *A.R.C. Meat Research Institute, Langford, Bristol BS18 7DY, UK*

*Present address: S. A. Frigorifico Monte Grande, San Martin 432/525, Monte Grande, Buenos Aires, Argentina

M. J. M. MICHELS, *Department of Microbiology, Unilever Research Duiven, Zevenaar, The Netherlands*

G. J. MORRIS, *Institute of Terrestrial Ecology, Culture Centre of Algae and Protozoa, 36 Storey's Way, Cambridge CB3 ODT, UK*

C. K. MURRAY, *Torry Research Station, 135 Abbey Road, Aberdeen AB9 8DG, UK*

D. R. ORR, *River Laboratory, Freshwater Biological Association, East Stoke, Wareham, Dorset BH20 BB6, UK*

B. REITER, *National Institute for Research in Dairying, University of Reading, Shinfield, Reading RG2 9AT, UK*

M. ELISABETH SHARPE, *National Institute for Research in Dairying, University of Reading, Shinfield, Reading RG2 9AT, UK*

B. G. SHAW, *A.R.C. Meat Research Institute, Langford, Bristol BS18 7DY, UK*

J. M. SHEWAN, *Torry Research Station, 135 Abbey Road, Aberdeen AB9 8DG, UK*

I. VANCE,* *Dunstaffnage Marine Research Laboratory, P.O. Box 3, Oban, Argyll PA34 4AD, UK*

D. D. WYNN-WILLIAMS, *Life Sciences Division, Bristol Antarctic Survey, Madingley Road, Cambridge CB3 OET, UK*

*Present address: Department of Life Sciences, Polytechnic of Central London, London W1M 8JS, UK

Preface

This volume, Number 13 in the Society for Applied Bacteriology Technical Series, is based on the demonstrations made at the Autumn 1977 Demonstration Meeting of the Society.

The subject of the meeting was "Cold Tolerant Microbes in Spoilage and the Environment". The book brings together contributions from experts on psychrotrophic bacteria and contains chapters on various aspects of low temperature microbiology as they affect fundamental metabolism, ecology and food spoilage. Providing as it does an excellent survey of the importance of psychrotrophic bacteria it will be of value generally to bacteriologists and is essential reading for those engaged in this particular field.

We would like to thank those who presented demonstrations at this meeting, members of the staff of the Department of Biology, Brunel University, especially Professor J. D. Gillett, Mr. F. Jones (Senior Lecturer) and Mrs. S. Bannerman (Chief Technician) who helped in the staging of this demonstration, and Mr. D. W. Lovelock of Heinz Co. Ltd., for his help and advice.

February 1979

A. D. Russell
R. Fuller

Contents

Microbial Growth at Low Temperatures

R. A. Herbert and M. Bhakoo

Department of Biological Sciences, The University, Dundee, Scotland

Introduction

Temperature is undoubtedly one of the most important environmental parameters operating in any ecosystem. The fundamental role of temperature in controlling the rate of all physico-chemical reactions underlies its significance when considering the complexities of biological processes. Micro-organisms, since they can be cultivated with relative ease in the laboratory, provide a convenient experimental tool for elucidating the effects of temperature on living cells. Most micro-organisms can grow only over a relatively narrow temperature range, from about −10°C to +70°C, and within this range, temperature affects the growth rate, nutritional requirements, enzyme activities and chemical composition of the cells. Considering the profound effects of temperature, it is somewhat disturbing that many microbiologists conveniently neglect the fact that most natural environments are at low temperatures. The literature is, however, replete with examples of micro-organisms whose physiology was studied at temperatures in excess of those from which they were obtained.

Traditionally, micro-organisms have been classified on the basis of temperature optima for growth into thermophiles, mesophiles and psychrophiles according to their ability to grow at high, medium and low temperatures, respectively. The first two groups can be readily differentiated by their growth temperature optima. Psychrophiles have also been defined following their discovery by Forster (1887) but it is only during the last decade that some semblance of agreement has been achieved amongst microbiologists in defining these organisms. For detailed discussions on the definition of psychrophiles the reader should consult reviews by Ingraham and Stokes (1959) and Morita (1966, 1975). For the purpose of this review the definition of psychrophiles as proposed by Morita (1975) will be used, i.e. organisms having an optimal growth temperature of about 15°C or lower, a maximal growth temperature of about 20°C and a minimal growth temperature of 0°C or lower. Bacteria which grow at 0°C and at maximum temperatures exceeding 25°C will be considered to be psychrotrophic.

Distribution of Psychrophilic Bacteria

The ability of micro-organisms, and in particular bacteria, to grow at low temperatures was first demonstrated by Forster (1887) and subsequently they have been shown to have a widespread distribution in a range of natural environments (Ingraham and Stokes, 1959; Morita, 1966). The majority of the isolates obtained by the early investigators were psychrotrophic and the occurrence of truly psychrophilic bacteria, which conform to the present definition, were not reported until 1964. Morita and Haight (1964) isolated the first authentic psychrophilic bacterium, *Vibrio marinus* MP-1, and subsequently they have been isolated from a range of habitats (Harder and Veldkamp 1967; Stanley and Rose, 1967; Moiroud and Gounod, 1969; Herbert and Bell, 1977). The principal reason for the inability of early microbiologists to isolate psychrophiles was the failure to use pre-cooled media and to ensure that the samples were never exposed to lethal temperatures. These organisms are abnormally thermolabile and even exposure to room temperature for a period of time is likely to be lethal. However, if pre-cooled media are used, psychrophiles can be obtained relatively easily from permanently cold environments. The majority of psychrophiles studied have been isolated from the marine environment but they are not confined to this habitat.

Taxonomy of Psychrophilic Bacteria

A notable feature of psychrophilic bacteria is the predominance of Gram negative species isolated and the relative rarity of Gram positive organisms. Most of the psychrophiles studied belong to the genera *Vibrio* and *Pseudomonas* (Morita and Haight, 1964; Harder and Veldkamp, 1967; Herbert and Bell, 1977). Chromogenic isolates have been assigned to the genus *Flavobacterium* (Stanley and Rose, 1967; Sieburth, 1967), although the taxonomic status of the genus is at present uncertain. Gram positive psychrophiles have been isolated. Sieburth (1967) isolated *Arthrobacter* spp. from Narragansett Bay, Rhode Island and a similar organism, *A. glacialis*, has been isolated from sediments below arctic glaciers (Moiroud and Gounod, 1969). Until recently there had been no reports of psychrophilic anaerobes. Sinclair and Stokes (1963) successfully isolated from soil, mud and sewage *Clostridium* spp. which were capable of sporulating at 0°C, but these were shown to be psychrotrophs. Liston *et al.* (1969) and then Finnes and Matches (1974) were the first to report psychrophilic *Clostridia* in marine sediments from Puget Sound. Sixteen of these isolates had growth temperature maxima below 15°C.

Herbert and Bell (1973, 1974) made an extensive survey of the waters and sediments of the lakes and offshore coastal waters for psychrophilic representatives of other bacterial groups. Table 1 shows the diversity of bacterial types that were isolated. The majority of the isolates obtained were psychrotrophic. Although psychrophilic representatives of some groups have been obtained they are present in relatively small numbers. It seems paradoxical that in such permanently cold-environments (lake temperature maximum 5°C, sea temperature maxima −1°C) that a greater proportion of the microflora is not psychrophilic.

TABLE 1. Bacterial groups isolated from freshwater lakes and coastal offshore waters at Signy Island, South Orkney Islands, Antarctica

Bacterial group	Marine	Freshwater
Azotobacter sp.	+	+
Nitrosomonas sp.	—	+
Nitrobacter sp.	—	+
Dentrifier sp.	+	+
Proteolytic sp.	+	+
Purple non-sulphur sp.	+	+
Purple sulphurs	+	+
Green sulphurs	+	+
Thiobacillus sp.	+	+
Filamentous S-oxidizers (*Beggiatoa*)	ND	+
Sulphate reducers	+	+
Sulphur oxidizers	+	ND
Cellulose decomposers	+	+

ND, not determined.

Growth of Psychrophiles

The growth characteristics of psychrophilic bacteria have been studied extensively over the last decade. Unfortunately, many of these studies have been performed on batch grown cultures which make interpretation of the data difficult. Our data, like those of Harder and Veldkamp (1967), relate to psychrophilic bacteria grown in a chemostat. In this way, by growing the cells at a constant growth rate, the effect of temperature can be studied as a single environmental factor without the complication of

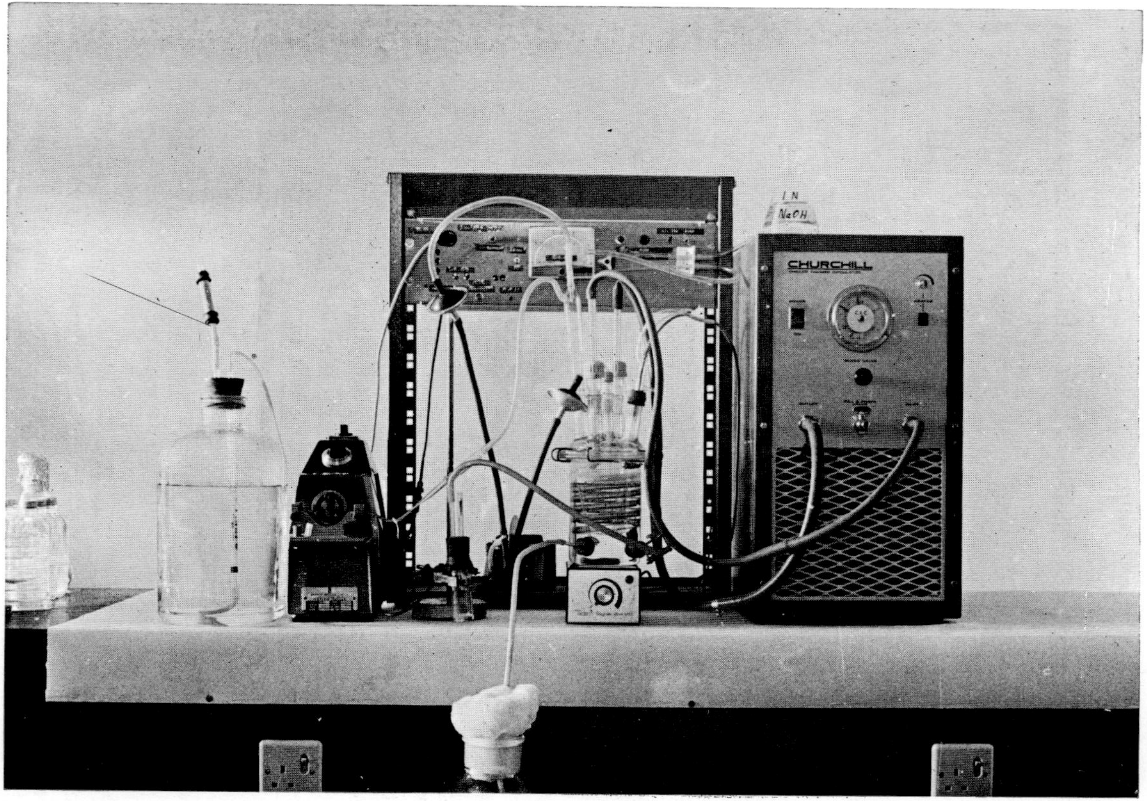

FIG. 1. Chemostat arrangement for growing psychrophilic bacteria

changes in growth rate, nutrient concentration, pH, dissolved gas concentrations which occur during the growth of bacteria in batch culture. In our own work we have used a simple single stage 1 litre chemostat (Fig. 1) based on the design of Baker (1968). Cooling is provided from an external thermocirculator (Churchill Instruments Ltd) circulating coolant via a "cold-finger" or cooling coil in the chemostat. In order to standardize the growth conditions we have used a defined mineral salts medium (Brown and Stanley, 1972) supplemented with glucose as carbon source and either KNO_3 or NH_4Cl as nitrogen source.

Temperature Effects

Effect of temperature on growth rate

It is a widely held belief that psychrophilic bacteria grow slowly at near zero temperatures. A closer examination of the literature shows that this is not necessarily the case. *Vibrio marinus* MP-1, for example, has a mean generation time of 226 min at 3°C (Morita and Albright, 1965) whilst Larkin and Stokes (1969) claim that *Bacillus cryophilus* will double its population every 6 h at –5°C. Other psychrophilic bacteria are much slower growing. *Micrococcus cryophilus* has a doubling time of 28·33 h at 0°C and *Vibrio* AF-1 (Herbert and Bell, 1977) has a mean generation time of 23 h. It may well be that it is not the rate of growth which is important but the efficient conversion of substrate carbon into cell carbon which is significant.

Following the derivation by Arrhenius (1889) of an equation to describe the effects of temperature on chemical reactions,

$$K = Ae^{-E/RT} \qquad (1)$$

where K is the reaction rate, R the gas constant, T the absolute temperature, $E(\mu)$ the activation energy and A a constant, many microbiologists have attempted to apply the equation to the growth of bacterial cultures. By substituting bacterial growth rate for K in the equation the temperature characteristic for growth (μ) can be determined. Ingraham (1958, 1962) originally proposed that this could be used to determine whether or not a particular micro-organism was a psychrophile or a mesophile since the former should have a lower μ value. Ingraham's data have been challenged by Janota-Bassalik (1963) and Hanus and Morita (1968) on the grounds that his calculations were erroneous. The latter authors could find no significant difference between the μ values of a psychrophile, a psychrotroph and a mesophile. Similar findings have been reported for

TABLE 2. Optimal growth temperatures and temperature characteristic of growth (μ) for 5 psychrophilic *Vibrio* spp.

Isolate number	Optimal growth temperature (0°C)	Temperature characteristic for growth
AM1	6	27 400
AM10	10	32 000
AF1	15	9900
BM2	4	8200
BM4	8	38 700

yeasts (Shaw, 1967) and Gram positive bacteria (Brownlie, 1966). We determined μ values for 5 psychrophilic *Vibrio* spp. (Table 2) and these similarly show no consistent values. Whilst μ values may be of doubtful validity in describing the growth characteristics of psychrophiles and mesophiles, the Arrhenius plot of log specific growth rate against $^1/_{°K}$ is significant. In the case of psychrophiles, the slope of the Arrhenius plot is linear down to 0°C and below, whilst psychrotrophs deviate from linearity about 4–5°C and mesophiles tend to deviate from a straight line at higher temperatures (Harder and Veldkamp, 1971).

Effect of temperature upon substrate uptake

The theory that permeability and the associated control of solute transport may constitute a major factor in the ability of psychrophilic microorganisms to grow at low temperatures has gained considerable support and evidence in recent years. The minimum temperature for the growth of mesophiles is considered by many authors to be controlled by the low temperature inhibition of substrate uptake (Ingraham and Bailey, 1959; Rose and Evison, 1965; Morita and Buck, 1974). Even in a psychrotrophic *Vibrio*, Paul and Morita (1971) found that uptake of ^{14}C-glutamate was much reduced at low temperatures. However, Baxter and Gibbons (1962) and Cirillo *et al.* (1963) showed that in a psychrophilic *Candida* sp. sugar transport was largely independent of temperature. These data have been subsequently confirmed for Gram positive and negative bacteria (Wilkins, 1973). In *M. cryophilus* uptake of lysine occurred at the same rate when the cells were grown at 0°C as at 20°C (Russell, 1971). Herbert and Bell (1977) showed that in *Vibrio* AF-1 maximum uptake of ^{14}C-glucose and lactose occurred at 0°C and decreased with increasing temperature. Using statistical analysis (analysis of variance) these authors found that the temperature of assay and not the temperature at which the cells were grown was the factor controlling the rate of sugar uptake, i.e. there appear to be no quantitative differences in the number of glucose or lactose permease molecules present in the cells when grown at different temperatures.

Closely allied to the effects of temperature on substrate uptake are the effects on the composition of the membrane lipids. Micro-organisms, in common with most other poikilothermic and many homeothermic organisms, synthesize increased proportions of unsaturated fatty acids, at the expense of saturated acids, when the growth temperature is lowered (Farrell and Rose, 1967). It is well known that an increase in the degree of unsaturation in lipids causes a decrease in the melting point of the lipids. A number of workers have argued that the physiological effect of the increased synthesis of unsaturated fatty acids at low temperatures is to maintain the membrane lipids in a fluid, and therefore mobile, state. This "lipid solidification" theory has been difficult to demonstrate experimentally. Circumstantial evidence for the theory has been provided by studies on cold shock in mesophilic and psychrophilic pseudomonads (Farrell and Rose, 1968). *Pseudomonas aeruginosa*, a mesophile, when grown at 30°C is susceptible to cold shock and loses viability whereas when grown at 10°C the cells are no longer susceptible. At the lower temperature (10°C) increased levels of unsaturated fatty acids are synthesized and Farrell and Rose (1968) postulate that these prevent lesions forming in the cytoplasmic membrane, thus maintaining their integrity. Additional evidence to support this hypothesis was the demonstration that a psychrotrophic pseudomonad which had increased unsaturated fatty acid levels at 30°C, was less susceptible to cold shock than the mesophile at this temperature. These data provide evidence that solute transport may well be influenced by the degree of unsaturation of the fatty acid side chains of the membrane lipids.

Several studies have been made of the changes in fatty acid composition of membrane lipids with temperature; however, the interpretation of certain of these data is difficult since the organisms were grown in batch culture and it is known that fatty acid composition alters in respect of growth rate as well as temperature. Comparative studies of mesophilic and psychrophilic yeasts have shown that the psychrophiles are generally endowed with a higher proportion of unsaturated fatty acids than the mesophiles (Kates and Baxter, 1962; Brown and Rose, 1969a). Data on psychrophilic bacteria have not provided such clear-cut findings. Kates and Hagen (1964) examined mesophilic and psychrophilic *Serratia* spp. and concluded that the psychrophilic species produced increased unsaturated fatty acids (hexadecenoic and octadecenoic acids) compared with the mesophiles. In *M. cryophilus* the response to low temperature is somewhat different and involves a shortening of the fatty acid chain length rather than increased synthesis of unsaturated fatty acids (Russell,

1971) whilst C. N. Brown and Minnikin (pers. comm.) found no change in the fatty acid composition of several psychrophilic marine pseudomonads grown in continuous culture at different temperatures. We have made a detailed study of the fatty acid and phospholipid composition of 4 psychrophilic *Vibrio* spp. (Bhakoo and Herbert, in preparation). Data for *Vibrio* AF-1 are presented in Tables 3 and 4. As

TABLE 3. Effect of temperature on the free fatty acid composition of *Vibrio* AF-1 grown on glucose under carbon limitation in a chemostat (D 0·02 h^{-1})

Fatty acid[a]	Temperature of growth (°C) 0	8	15
C_8:0	3·44	3·39	0·44
C_9:0	0·51	0·21	2·34
C_9:1	0·22	0·21	—
C_{10}:0	0·23	2·15	7·16
C_{11}:0	—	1·37	1·98
C_{12}:0	0·22	1·42	0·43
C_{12}:1	1·31	0·06	—
C_{13}:0	1·94	0·46	—
C_{13}:1	0·85	0·34	1·35
C_{14}:0	3·78	7·67	5·92
C_{14}:1	4·49	2·57	1·1
C_{15}:0	10·31	1·59	6·33
C_{15}:1	14·18	9·15	9·90
C_{16}:0	5·81	10·32	11·81
C_{16}:1	4·50	0·67	3·57
C_{17}:0	—	2·17	4·45
C_{17}:1	4·35	—	—
unsaturated / saturated	1·12	0·43	0·42
Total	56·85	43·2	56·87

[a] Units: $\mu g\ mg^{-1}$ dry wt.

the growth temperature is lowered there is a switch to the production of unsaturated fatty acids (C_{12}:1, C_{14}:1, C_{15}:1, C_{16}:1 and C_{17}:1). In addition there is evidence of chain shortening also occurring with the appearance of C_8:0, C_9:1, C_{12}:1 and C_{13}:0 fatty acids. *Vibrio* AF-1 thus appears to have a dual response to decreasing temperature in respect of its fatty acid composition. Changes also occur in the phospholipid composition of this bacterium (Table 4). The total quantities of phospholipid are significantly greater at 0°C than at either 8 or 15°C. Quantitatively, there are increases in phosphatidylserine, diphosphotidyglycerol and a com-

TABLE 4. Effect of temperature on the phospholipid composition of *Vibrio* AF-1 grown on glucose under carbon limitation in a chemostat (D 0·02 h^{-1})

Phospholipid[a]	Temperature (°C) 0	8	15	20
Phosphatidylserine	48·92	39·06	28·03	21·68
Phosphatidylglycerol	36·34	42·15	52·91	34·72
Phosphatidyethanolamine	26·23	24·49	18·09	17·69
Cardiolipin	20·16	15·02	15·14	14·68
Lysophosphatidylglycerol[b]	11·56	7·42	6·72	2·11
Total phospholipid	143·21	128·14	120·87	90·88

[a] Units: μg mg^{-1} dry wt.
[b] Tentatively identified as this phospholipid.

ponent, tentatively identified as lysophosphatidylglycerol. Similar data have been obtained for the other psychrophilic *Vibrio* spp. that we have examined. Oliver and Colwell (1973) also noted changes in the phospholipid composition of the psychrotroph *Vibrio marinus* PS-207 at low temperatures, particularly an increase in lysophosphatidylglycerol, lysophosphatidylethanolamine and diphosphatidylglycerol. Other workers, however, report no changes in phospholipid composition with temperature in the Gram negative bacteria they have studied. Nevertheless, there is now good evidence to show that changes do occur in the fatty acid and phospholipid composition of psychrophilic bacteria in response to decreased temperatures but the significance of these findings has yet to be elucidated.

Effect of temperature on the macromolecular composition of psychrophiles

It has been postulated that in order to compensate for growth at low temperatures many psychrophilic bacteria synthesize increased quantities of enzymes. Indirect evidence to support this contention comes from analyses of the nucleic acid and protein contents of psychrophilic bacteria at low temperatures (0°C). Harder and Veldkamp (1967) showed that their psychrophilic pseudomonad synthesized 40% more ribonucleic acid (RNA) and 11–14% more protein at 0°C than at 14°C, the optimal growth temperature for this isolate. Data presented in Fig. 2 show that a similar increase in RNA and protein contents occurs with *Vibrio* AF-1. At 15°C, the isolate's optimal growth temperature, RNA and protein content reach a minimum but at 0°C have increased by 29% and 13%, respectively. Microscopical examination of the cells grown at 15°C and 0°C showed that they were significantly larger (Table 5) at the lower temperature and cell volumes had increased by approximately 43%.

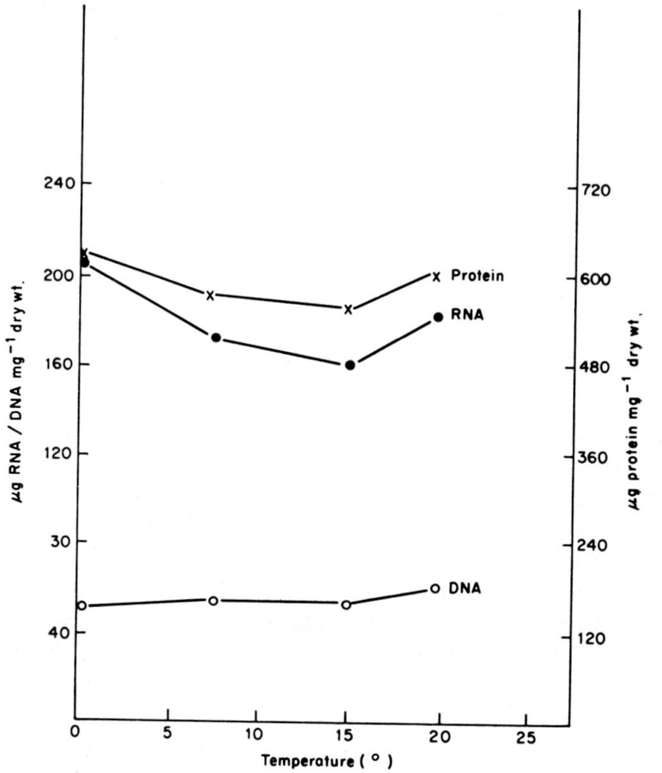

FIG. 2. Effect of growth temperature on the macromolecular composition of *Vibrio* AF-1 cells grown on glucose under carbon limitation in a chemostat (D 0·02 h^{-1}).

Similar data have been reported by Brown and Rose (1969b) for the yeast *Candida utilis* NCYC 321 when grown at 15°C. In *Klebsiella aerogenes* (Tempest and Hunter, 1965) and a psychrophilic marine pseudomonad (Harder and Veldkamp, 1967) the majority of the RNA synthesized at low temperatures is ribosomal RNA. These data support indirectly the contention that increased enzyme synthesis occurs at low temperatures to compensate for reduced enzyme activity.

Effect of temperature on respiratory activity and cell yield

Several studies have investigated the effects of temperature on the respiratory activity of psychrophilic micro-organisms when grown at near zero temperatures. Harder and Veldkamp (1967) demonstrated with

TABLE 5. Effect of temperature on cell numbers, cell size and cell volume of *Vibrio* AF-1 when grown on glucose under carbon limitation in a chemostat (D 0·02 h^{-1})

Growth temperature (°C)	Cell length (μm)	Cell width (μm)	Cell volume (μ^3)	Cell nos ml^{-1}
0	2·15	0·67	0·76	3·15 × 10^9
8	1·67	0·74		3·7 × 10^9
15	1·52	0·66	0·53	1·5 × 10^9
20	—[a]	—[a]	—[a]	—[a]

[a] No measurements (cells forming filaments).

their psychrophilic marine pseudomonad grown in continuous culture that O_2 consumption was minimal at the organism's optimal growth temperature (14°C) and increased markedly at both sub-optimal and supra-optimal growth temperatures. These workers also showed the intimate relationship between cell yield and O_2 consumption. At 14°C, the pseudomonad's optimal growth temperature, cell yield was maximum and O_2 consumption minimum (42 μl mg^{-1} dry wt. h^{-1}) whereas at 5°C O_2 consumption had increased to 60 μl mg^{-1} dry wt. h^{-1} and cell yield had declined. Harder and Veldkamp (1967) interpreted these data as demonstrating an increase in energy-generating enzymes in an attempt to compensate for the reduced metabolic activity occurring at sub-optimal growth temperatures thereby maintaining growth rate at the expense of cell yield. Experiments with *Vibrio* AF-1 grown under carbon limitation in a chemostat show a similar pattern to those observed by Harder and Veldkamp (1967) for their marine pseudomonad Minimum O_2 consumption occurs, however, not at the optimal growth temperature (15°C) but at 0–8°C, the temperature of maximum cell yield (Table 6). Determination of RQ values by Warburg manometry of washed cell suspensions of *Vibrio* AF-1 show that at 0°C the RQ is < 1 and substrate carbon is available for biosynthesis. As the temperature is increased the RQ approaches unity suggesting that most of the carbon assimilated is oxidized to yield energy rather than being channelled into biosynthesis. We have attempted to study this further by determining the "maintenance energy" of *Vibrio* AF-1 at different temperatures.

The concept of "maintenance energy" was first introduced by Pirt (1965) to account for the reduction in the theoretical growth yield at low dilution rates. The theory predicts that in the production of a certain cell mass a given quantity of the substrate is used for new cell material and a further quantity for maintenance of the existing cell population. Pirt (1965) developed a relationship between cell yield and growth rate in terms of a maintenance coefficient and expressed this as the amount of substrate utilized per unit time per cell for maintenance purposes.

TABLE 6. Effect of temperature on the RQ oxygen consumption, and maintenance coefficient of *Vibrio* AF-1 when grown on glucose under carbon limitation in a chemostat (D 0·020 h^{-1})

Growth temperature (°C)	Q O_2[a]	Q CO_2[a]	RQ	Cell yield[b]	Maintenance coefficient[c]
0	34	28	0·82	0·65	0·021
8	30	28	0·93	0·56	0·042
15	54	50	0·93	0·34	0·080
20	74	74	1·0	0·17	0·096

[a] μl mg^{-1} dry wt. h^{-1}.
[b] mg dry wt. mg^{-1} carbon sources.
[c] g glucose g^{-1} dry wt. h^{-1}.

Harder and Veldkamp (1967) argued from their data that there was an increase in energy of maintenance at temperatures either side of the optimum for growth of their marine pseudomonad but were unable to demonstrate this unequivocally. Data obtained for *Vibrio* AF-1 (Table 6) provides further experimental evidence to support the view of Harder and Veldkamp (1967). At 0°C the maintenance chemostat coefficient for *Vibrio* AF-1 when grown on glucose under carbon limitation in a chemostat is approximately 2% of the carbon input whereas at 20°C this has increased to 10% of the carbon input.

In terms of cell yield, there appears to be considerable variation depending upon the organisms studied. Harder and Veldkamp (1967) obtained a maximum specific growth rate of 0·20 at 16°C with their *Pseudomonas* sp. yet the cell yield was only 0·35 mg dry wt. mg^{-1} carbon utilized. *Vibrio* AF-1 in contrast grows much more slowly (μ_{max} 0·05 at 15°C) but is very much more efficient in its use of external carbon. Under glucose limitation a yield coefficient of 0·60 (mg dry wt. cells mg^{-1} carbon utilized) was obtained (Table 6). This high yield is comparable to that obtained for *Escherichia coli* when grown aerobically on glucose at 37°C (Hernandez and Johnson, 1967). It may well be that more rapidly growing psychrophilic bacteria, such as *V. marinus* MP-1 and *Pseudomonas* spp. sacrifice cell yield for a more rapid rate of growth whereas *Vibrio* AF-1, *M. cryophilus* and related species may gain no competitive advantage by growing rapidly at low temperatures but have adapted by utilizing efficiently the carbon available in the habitats from which they were isolated. Data presented in Fig. 3 show the main physiological features of *Vibrio* AF-1 and how this organism appears to be well adapted for growth at near zero temperatures. However, much more data are required before a full explanation of growth at low temperatures can be made.

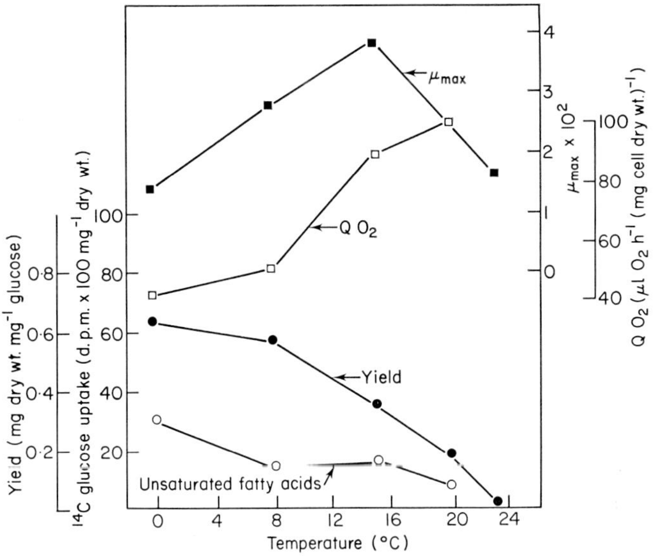

FIG. 3. The effect of temperature on the physiology of *Vibrio* AF-1 cells grown on glucose under carbon limitation in a chemostat (D 0·02 h^{-1}).

Significance of Psychrophiles in Natural Environments

Whilst psychrophilic bacteria have been isolated from a wide range of natural environments (for further information see reviews by Ingraham and Stokes, 1959; Morita, 1966, 1975) very few studies have been made to assess the significance of these micro-organisms in aquatic and terrestrial ecosystems. Probably the most extensive and detailed study was made by Sieburth (1967) who investigated the effect of water temperature, on a seasonal basis, on the heterotrophic bacterial flora of Narragansett Bay, Rhode Island. During winter, when water temperatures reach a minimum (—2°C), psychrophilic bacteria became the dominant component of the microflora whereas during the summer, when water temperatures reach +23°C (August), psychrotrophic and mesophilic species became dominant. These field data were confirmed when a more detailed investigation was made of selected isolates obtained during the year. Sieburth (1967) noted that the effect of temperature was evident irrespective of the taxonomic status of the isolates and that he could find no evidence for the enhancement or suppression of one taxonomic group as a result of temperature. Similar data (Tajima *et al.*, 1974) have been reported for a study of seasonal changes in bacterial populations in off-shore waters of Hakodata Bay, Japan. Whilst these studies do indicate that populations of psychrophilic bacteria do develop at particular times of the year in temperate climates, the significance of these data in relation to the actual or potential microbial activity are less well known. Harder and Veldkamp (1971) in an elegant study using mixed populations of psychrophilic and psychrotrophic bacteria grown at several temperatures in a chemostat showed that −2°C the psychrophile outgrew the psychrotroph at all dilution rates whereas at 16°C the reverse situation occurred. At intermediate temperatures of 4 and 10°C only the dilution rate influenced which bacterium was dominant. The psychrophile predominated at high dilution rates whereas the psychrotroph grew better and predominated at low dilution rates. From this study it is apparent that temperature does apply a selective pressure and the organism which predominates is that which has thermal growth characteristics most closely fitting the environmental conditions. If such data are assumed to be generally applicable then it becomes apparent that in permanently cold environments such as ocean waters and the polar regions that psychrophilic bacteria will largely be responsible for nutrient cycling. Such a contention would be especially pertinent in the surface waters of the Antarctic Ocean so rich in primary nutrients. So far this has not been demonstrated.

Acknowledgements

Part of this work was made possible by Grant GR/3/1836 from the Natural Environmental Research Council. M. Bhakoo gratefully acknowledges the receipt of a SRC Research Studentship.

References

ARRHENIUS, S. (1889). Cited by Farrell, J. & Rose, A. H. 1967. Temperature effects on micro-organisms. In *Thermobiology* (Rose, A. H., ed.). London and New York: Academic Press.

BAKER, K. (1968). Low cost continuous culture apparatus. *Laboratory Practice* **17,** 817–821.

BAXTER, R. M. & GIBBONS, N. E. (1962). Observations on the physiology of psychrophilism in a yeast. *Canadian Journal of Microbiology* **8,** 511–517.

BROWN, C. M. & ROSE, A. H. (1969a). Fatty acid composition of *Candida utilis* as affected by growth temperature and dissolved oxygen tension. *Journal of Bacteriology* **99,** 371–378.

BROWN, C. M. & ROSE, A. H. (1969b). Effect of temperature on composition and cell volume of *Candida utilus*. *Journal of Bacteriology* **97,** 261–272.

BROWN, C. M. & STANLEY, S. O. (1972). Environment mediated changes in the cellular content of the "pool" constituents and their associated changes in cell physiology. *Journal of Applied Chemistry and Biotechnology* **22,** 363–389.

BROWNLIE, L. E. (1966). Effects of some environmental factors on psychrophilic *Microbacteria. Journal of Applied Bacteriology* **29,** 447–454.

CIRILLO, V. P., WILKINS, P. O. & ANTON, J. (1963). Sugar transport in a psychrophilic yeast. *Journal of Bacteriology* **86,** 1259–1264.

FARRELL, J. & ROSE, A. H. (1967). Temperature effects on micro-organisms. In *Thermobiology* (Rose, A. H., ed.). London and New York: Academic Press.

FARRELL, J. & ROSE, A. H. (1968). Cold shock in a mesophilic and a psychrophilic pseudomonad. *Journal of General Microbiology* **50,** 429–439.

FINNES, G. & MATCHES, J. R. (1974). Low temperature growing clostridia from marine sediments. *Canadian Journal of Microbiology* **20,** 1639–1645.

FORSTER, J. (1887). Ueber einige Eigenschaften leuchtender Bakterien. *Centr Bakteriel Parasitenk* **2,** 337–340.

HANUS, F. J. & MORITA, R. Y. (1968). Significance of the temperature characteristic for growth. *Journal of Bacteriology* **95,** 736–737.

HARDER, W. & VELDKAMP, H. (1967). A continuous culture study of an obligately psychrophilic *Pseudomonas* sp. *Archïv für Mikrobiologie* **59,** 123–130.

HARDER, W. & VELDKAMP, H. (1971). Competition of marine psychrophilic bacteria at low temperature. *Antonie van Leeuwenhoek* **37,** 51–63.

HERBERT, R. A. & BELL, C. R. (1973). Nutrient cycling in freshwater lakes on Signy Island, South Orkney Islands. *British Antarctic Survey Bulletin* **37,** 15–20.

HERBERT, R. A. & BELL, C. R. (1974). Nutrient cycling in the Antarctic marine environment. *British Antarctic Survey Bulletin* **39,** 7–11.

HERBERT, R. A. & BELL, C. R. (1977). Growth characteristics of an obligately psychrophilic *Vibrio* sp. *Archives of Microbiology* **113,** 215–220.

HERNANDEZ, E. & JOHNSON, M. J. (1967). Energy supply and cell yield in aerobically grown microorganisms. *Journal of Bacteriology* **94,** 996–1001.

INGRAHAM, J. L. (1958). Growth of psychrophilic bacteria. *Journal of Bacteriology* **76,** 75–80.

INGRAHAM, J. L. (1962). Newer concepts of psychrophilic bacteria. In *Proceedings of Low Temperature Microbiology Symposium* 1961. Camden, New Jersey: Campbell Soup Company, pp. 41–56.

INGRAHAM, J. L. & BAILEY, G. F. (1959). Comparative effect of temperature on the metabolism of mesophilic and psychrophilic bacteria. *Journal of Bacteriology* **77,** 609–613.

INGRAHAM, J. L. & STOKES, J. L. (1959). Psychrophilic bacteria. *Bacteriological Reviews* **23,** 97–108.

JANOTA-BASSALIK, L. (1963). Growth of psychrophilic and mesophilic strains of peat bacteria. *Acta microbiologica Polanica* **12,** 41–45.

KATES, M. & BAXTER, R. M. (1962). Lipid composition of mesophilic and psychrophilic yeasts as influenced by environmental temperatures. *Canadian Journal of Biochemistry and Physiology* **40,** 1213–1227.

KATES, M. & HAGEN, P. O. (1964). Influence of temperature on fatty acid composition of mesophilic and psychrophilic *Serratia* sp. *Canadian Journal of Biochemistry* **42,** 481–488.

LARKIN, J. M. & STOKES, J. L. (1969). Growth of psychrophilic microorganisms at sub-zero temperatures. *Canadian Journal of Microbiology* **14,** 97–101.

LISTON, J., HOLMAN, M. & MATCHES, J. (1969). Psychrophilic clostridia from marine sediments. *Bacteriological Proceedings,* 35–40.

MOIROUD, A. & GOUNOT, A. M. (1969). Sur une bacteria psychrophile obligatoire isolee de limins glaciaires. *Comptes Rendus Academie Science Series D* **269,** 2150–2152.

MORITA, R. Y. (1966). Marine psychrophilic bacteria. *Oceanography and Marine Biology Annual Review* **4,** 105–121.

MORITA, R. Y. (1975). Psychrophilic bacteria. *Bacteriological Reviews* **39,** 146–167.

MORITA, R. Y. & ALBRIGHT, L. J. (1965). Cell yields of *Vibrio marinus*, an obligate psychrophile, at low temperature. *Canadian Journal of Microbiology* **11,** 221–227.

MORITA, R. Y. & BUCK, G. E. (1974). Low temperature inhibition of substrate uptake. In *Effect of the ocean environment on microbial activities* (Colwell, R. R. & Morita, R. Y., eds). Baltimore: University Park Press, pp. 124–129.

MORITA, R. Y. & HAIGHT, R. D. (1964). Temperature effects on the growth of an obligately psychrophilic marine bacterium. *Limnology and Oceanography* **9,** 103–106.

OLIVER, J. D. & COLWELL, R. R. (1973). Extractable lipids of Gram negative bacteria. Phospholipid composition. *Journal of Bacteriology* **114,** 897–908.

PAUL, K. L. & MORITA, R. Y. (1971). Effects of hydrostatic pressure and temperature on the uptake and respiration of amino acids by a facultatively psychrophilic marine bacterium. *Journal of Bacteriology* **108,** 835–843.

PIRT, S. J. (1965). The maintenance energy of bacteria in growing cultures. *Proceedings of the Royal Society, Series B* **163,** 224–231.

ROSE, A. H. & EVISON, L. M. (1965). Studies on the biochemical basis of minimum temperatures for growth of certain psychrophilic and mesophilic microorganisms. *Journal of General Microbiology* **38,** 131–141.

RUSSELL, N. J. (1971). Alteration in fatty acid chain length in *Micrococcus cryophilus* grown at different temperatures. *Biochimica et Biophysica Acta* **231,** 254–256.

SHAW, M. K. (1967). Formation of filaments and synthesis of macromolecules at temperatures below the minimum for the growth of *Escherichia coli*. *Journal of Bacteriology* **95,** 221–230.

SIEBURTH, J. McN. (1967). Seasonal selection of estuarine bacteria by water temperature. *Journal of Experimental Marine Biology and Ecology* **1,** 98–121.

SINCLAIR, N. A. & STOKES, J. L. (1963). Isolation of obligately anaerobic psychrophilic bacteria. *Journal of Bacteriology* **87,** 562–565.

STANLEY, S. O. & ROSE, A. H. (1967). Bacteria and yeasts from lakes on Deception Island. *Proceedings of the Royal Society, Series B* **252,** 199–207.

TAJIMA, K., DAIKU, K., EZURA, Y., KIMURA, T. & SAKAI, M. (1974). Procedure for the isolation of psychrophilic marine bacteria. In *Effect of the Ocean environment on Microbial Activities*. (Colwell, R. R. & Morita, R. Y., eds). Baltimore: University Park Press.

TEMPEST, D. W. & HUNTER, J. R. (1965). The influence of temperature and pH on the macromolecular composition of magnesium and glycerol limited *Aerobacter aerogenes* growing in a chemostat. *Journal of General Microbiology* **41,** 267–273.

WILKINS, P. O. (1973). Psychotrophic Gram positive bacteria: temperature effects on growth and solute uptake. *Canadian Journal of Microbiology* **19,** 909–915.

The Response of Unicellular Algae to Freezing and Thawing

G. J. Morris

Institute of Terrestrial Ecology, Culture Centre of Algae and Protozoa, Cambridge, Cambridgeshire, England

Introduction

The many variables which determine the response of cells to the stresses of freezing and thawing can be separated into two groups. First, there are the cellular factors, including the stage in the cell cycle, the age of culture and the growth temperature. Secondly, there are the physical determinants such as the rates of cooling and warming, the final temperature attained and the time of exposure to a sub-zero temperature. This paper deals with this second class of variables together with details of the equipment used in this laboratory to study them. Comprehensive reviews on cryobiology have been published elsewhere (Meryman, 1966; Mazur, 1970; MacLeod and Calcott, 1976).

Viability Assays

Many methods have been used to determine cell viability following freezing and thawing and these include motility, dye exclusion, uptake of fluorescein dyes (McGrath *et al.*, 1975; Towill and Mazur, 1975) and the ability to replicate. In most studies with micro-organisms, the ability to multiply is used since it has been shown that results obtained with indirect methods often do not correlate with growth measurements (Leibo *et al.*, 1970). With cells that form clumps it is necessary to make counts both before and after freezing because during freezing and thawing clumps of cells may break down into smaller units. This results in errors when survival is determined by colony formation (Morris, 1976). In order to compare the effects of freezing and thawing on different cell-types the median lethal temperature (LT_{50}) is defined as the temperature at which 50% of the cells are lost under standard cooling and warming conditions.

General Methods used in Freezing and Thawing

Ampoules

Glass containers can shatter when warmed from sub-zero temperatures, therefore polypropylene screw cap ampoules are now generally used. Polypropylene ampoules designed for low temperature work are commercially available (Sterilin, Nunc). For rapid rates of cooling, cells are placed in stainless steel hypodermic tubing or in glass capillaries.

Temperature measurement

For slow rates of cooling a 28 s.w.g. copper–constantan thermocouple was connected to a potentiometric recorder, whilst for faster cooling rates ($> 100°C\ min^{-1}$) a 45 s.w.g. thermocouple was attached to a display oscilloscope. Thermocouples are calibrated periodically using a platinum resistance thermometer. Rates of temperature change are usually expressed as $°C\ min^{-1}$; however, this is only useful if the temperature range over which this is measured is also defined. The problems of temperature measurement during cooling and warming are reviewed by Bald (1975).

Rates of cooling

Many devices have been described which regulate the rate of cooling during freezing. Commercially available equipment is designed for routine cryopreservation procedures and therefore not ideally suited to experimental work because only one cooling rate or temperature may be studied at a time. A review of some cooling rate equipment used in cryopreservation and the problems associated with its design has been published (Pegg *et al.*, 1973). In this laboratory, three methods of altering the rate of cooling and of attaining different final sub-zero temperatures have been used.

With all methods the ampules were first placed in a pre-cooled alcohol bath maintained at 1°C below the melting point of the experimental solution. After allowing the temperatures to equilibrate for 5 min, the samples were nucleated by touching the surface of the solution with the tip of a Pasteur pipette containing a frozen solution identical to that in the ampoule. With pathogenic organisms ice nucleation may be initiated mechanically by tapping the precooled ampoule. After the dissipation of the latent heat of fusion the ampoule is transferred to the cooling apparatus which has been adjusted to be at that moment the same temperature as the sample. If the solutions are not nucleated then they freeze spontaneously at different temperatures during cooling resulting in samples with different thermal histories; this may effect cell survival.

Low-temperature baths

Baths designed to maintain stable temperatures in the range 0 to —45°C are commercially available (Fryka, Grant). These baths are used normally for nucleating samples but can also be used for a cooling rate if after the dissipation of the latent heat of fusion the temperature control is set at a minimum and the bath allowed to cool. The rate of cooling is exponential but, providing the bath has the same volume of coolant each time, this is reproducible. However, only a limited range of cooling rates are obtainable and for the Fryka bath (Model KB300) these are in the region of 0·2°C min^{-1} (between —5 and —25°C). The results obtained with four species of algae cooled by this method are presented in Table 1.

TABLE 1. The response of 4 species of algae to freezing and thawing and to hypertonic solutions

Species	CCAP strain	LT_{50}(°C)	Recovery (%) following exposure to NaCl[a]
Chlorella protothecoides	211/7a	> —25	92
Chlorella emersonii	211/8h	—6·3	61
Euglena gracilis	1224/5z	—5·3	48
Chlamydomonas reinhardii	11/32c	—2·9	< 0·1

[a] Cells were exposed to 1 M NaCl for 5 min at 20°C, then resuspended.

These four strains have different susceptibilities to freezing injury at this rate of cooling. The recovery of *Chlorella protothecoides* is unaffected by freezing to temperatures of —25°C, whilst *Chlamydomonas reinhardii* is extremely sensitive to injury induced by freezing and thawing; the other two species have recoveries that are intermediate between these two extremes. At slow rates of cooling, large extracellular ice crystals form and this removal of liquid water as ice, produces hypertonic solutions to which the cells are exposed for long periods of time during cooling; this leads to cellular dehydration (Mazur, 1970). Injury to plant tissue culture cells following slow rates of cooling has been correlated with the cellular response to shrinkage and rehydration at 20°C (Towill and Mazur, 1976); a similar agreement has now been observed with algae (Table 1).

Temperature gradient bar

This is a brass rod 35 mm in diam., 1 m long, one end of which is maintained in liquid nitrogen with the other end in an ice bath or a low temperature alcohol bath. Holes are drilled out along the length of the bar (4 cm apart) to a sufficient depth to take polypropylene ampoules (12·5 mm diam.). Good thermal contact between the ampoules and the rod is achieved by placing 0·01 ml of methanol in each hole. Heat gain to the bar is reduced by a covering of insulation material (Armaflex). Once the bar is in equilibrium (2–3 h), the temperatures attained are illustrated in Fig. 1. This apparatus is used to study the effects of a number of

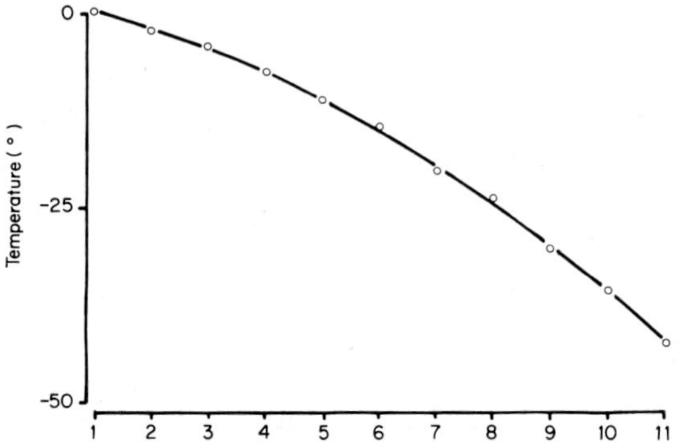

FIG. 1. Temperatures attained within a temperature gradient bar, one end of which was maintained at 0°C and the other end in liquid nitrogen.

different temperatures on the survival of organisms or for stepwise cooling.

Lagged cooling devices

These were originally described by Leibo *et al.* (1970). Samples were held equidistant from the centre of a circular aluminium holder to minimize variability between the cooling rates of individual tubes. The holder containing the ampoules is then placed in the interior of a freezing vessel whose exterior was cooled in liquid nitrogen. Industrial methylated spirit was used as a coolant within the vessel and was continuously stirred. Typical vessels used and the cooling rates achieved are given in Table 2. Very slow rates of cooling ($< 0{\cdot}1°C\ min^{-1}$) are obtained using silvered or strip silvered evacuated Dewars. The advantages of this method are that several cooling rates may be carried out simultaneously and that the rates of cooling achieved are reproducible.

Rapid cooling rates are achieved by placing ampoules directly into liquid nitrogen, or into primary cooling fluids such as Freon, isopentane or propane cooled in liquid nitrogen or a slurry of solid nitrogen prepared by the method of Sleytr (1970). Ultrarapid cooling rates can be obtained by capillary (Gehenio *et al.*, 1963) or spray (Plattner *et al.*, 1972) freezing small volumes of the sample.

TABLE 2. Typical lagged vessels and rates of cooling obtained

Vessel	Industrial methylated spirits (ml)	Cooling rate °C min⁻¹ −5 to −60°C
Evacuated, unsilvered Dewar (90 × 250 mm)	750	0·14
Evacuated, unsilvered Dewar (90 × 250 mm)	250	0·32
Unevacuated, unsilvered Dewar (90 × 250 mm)	1250	0·68
Unevacuated, unsilvered Dewar (90 × 250 mm)	1000	0·95
Unevacuated, unsilvered Dewar (90 × 250 mm)	750	1·40
Unevacuated, unsilvered Dewar (90 × 250 mm)	500	1·70
Unevacuated, unsilvered Dewar (90 × 250 mm)	250	2·75
Stainless steel beaker (100 × 165 mm)	1250	8
Stainless steel beaker (100 × 165 mm)	750	20
Stainless steel beaker (100 × 165 mm)	250	35

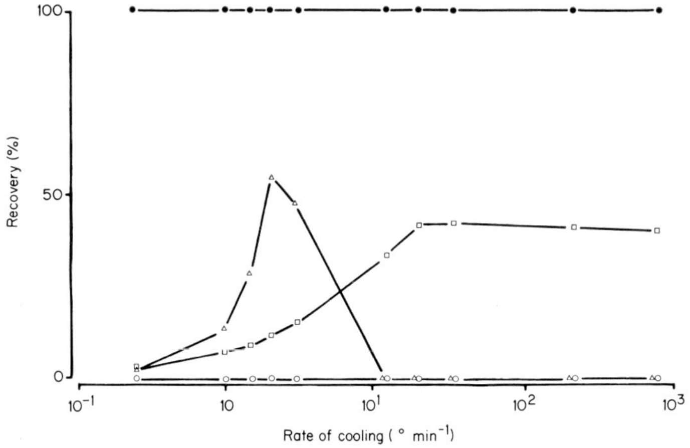

FIG. 2. Recovery (%) of *Chlorella protothecoides* (●), *Chlorella emersonii* (○), *Chlamydomonas nivalis* (△) and *Scenedesmus quadricauda* (□) after cooling at different rates to −196°C.

The effects of a range of cooling rates on the survival of four species of algae are given in Fig. 2. At slow rates, cells were cooled by lagged devices to –60°C then transferred to liquid nitrogen, for faster rates ampoules were immersed directly into different cooling fluids. *Chlorella protothecoides* has a recovery > 90%, whilst *Chlorella emersonii* and *Chlamydomonas reinhardii* had a survival of < 0·1% at all rates of cooling studied. An optimal rate of cooling was observed for *Chlamydomonas nivalis* and *Scenedesmus quadricauda* with damage increasing at both slower and faster rates of cooling. According to the "two-factor" theory of freezing injury (Mazur, 1970) cells cooled slower than the optimal rate are damaged by alterations in the properties of solutions induced by extracellular ice formation. At cooling rates faster than the optimal rate cells are injured by intracellular ice formation and its subsequent re-crystalization during warming.

Rates of warming

For most studies cells are warmed by rapid agitation of the ampoule in a water bath at 37°C. Slower rates of warming are achieved by using water baths at lower temperatures, allowing the ampoule to warm in the air or by using the cooling rate methods in reverse (Thorpe *et al.*, 1976).

References

BALD, W. B. (1975). A proposed method for specifying the temperature history of cells during the rapid cool-down of plant specimens. *Journal of Experimental Botany* **26,** 103–119.

GEHENIO, P. M., RAPATZ, G. & LUYET, B. (1963). Effects of freezing velocities in causing or preventing hemolysis. *Biodynamica* **9,** 78–82.

LEIBO, S. P., FARRANT, J., MAZUR, P., HANNA, H. G. & SMITH, L. H. (1970). Effects of freezing on marrow cell suspensions, interactions of cooling and warming rates in the presence of PVP, sucrose or glycerol. *Cryobiology* **6,** 315–332.

MACLEOD, R. A. & CALCOTT, P. H. (1976). Cold shock and freezing damage to microbes. In *The survival of vegetative microbes* (Gray, T. R. G. & Postgate, J. R., eds.). Cambridge: Cambridge University Press.

MAZUR, P. (1970). Cryobiology: the freezing of biological systems. *Science* **168,** 939–949.

MCGRATH, J. J., CRAVALHO, E. G. & HUGGINS, C. E. (1975). An experimental comparison of intracellular ice formation and freeze-thaw survival of He la S-3 cells. *Cryobiology* **12,** 540–550.

MERYMAN, H. T. (1966). *Cryobiology*. London and New York: Academic Press.

MORRIS, G. J. (1976). The cryopreservation of *Chlorella* 1. Interactions of rate of cooling, protective additive and warming rate. *Archives of Microbiology* **107,** 57–62.

Pegg, D. E., Hayes, A. R. & Kingston, R. E. (1973). Cooling equipment for use in cryopreservation. *Cryobiology* **10,** 271–281.

Plattner, H., Fischer, W. M., Schmitt, W. W. & Bachman, L. (1972). Freeze etching of cells without cryoprotectants. *Journal of Cell Biology* **53,** 116–126.

Sleytr, U. B. (1970). Fracture faces in intact cells and protoplasts of *Bacillus stearothermophilus*. A study by conventional freeze-etching and freeze-etching or corresponding moieties. *Protoplasma* **71,** 295–307.

Thorpe, P., Knight, S. C. & Farrant, J. (1976). Optimal conditions for the preservation of mouse lymph node cells in liquid nitrogen using cooling rate techniques. *Cryobiology* **13,** 126–133.

Towill, L. & Mazur, P. (1975). Studies on the reduction of 2, 3, 5 triphenyl tetrazolium chloride as a viability assay for plant tissue cultures. *Canadian Journal of Botany* **53,** 1097–1102.

Towill, L. & Mazur, P. (1976). Osmotic shrinkage as a factor in freezing injury in plant tissue cultures. *Plant Physiology* **57,** 290–297.

The Construction and Application of Temperature Gradient Incubators

J. H. BAKER and D. R. ORR

River Laboratory, Freshwater Biological Association, East Stoke, Wareham, Dorset, England

Introduction

A temperature gradient incubator may be defined as an appliance which maintains several different temperatures within known limits inside a single relatively small apparatus. This type of apparatus has also been called a thermal-gradient block (Battley, 1964), a polythermostat (Oppenheimer and Drost-Hansen, 1960; Palumbo *et al.*, 1967) and minor variations on these themes, but the term temperature gradient incubator (TGI), though somewhat lengthy, seems to describe the function of the apparatus most clearly and is therefore preferred. A TGI may be used to investigate the response to different temperatures of both living organisms and abiotic systems (Selwyn, 1961). Although the apparatus is essentially the same for both, the latter will not be discussed here.

The first TGI was invented to investigate the temperature preferences of animals (Herter, 1934). A series of separate incubators could not have been used for this purpose because it was necessary for the animal to move freely to the temperature of its choice. It was more than 20 years later that the first microbes (*Chlorella pyrenoidosa* and *Anacystis nidulans*) were grown in a TGI by Halldal and French (1956) and they appear to have developed their apparatus independently, unaware of the earlier work by zoologists. Moreover, Halldal and French, unlike Herter, could have obtained the answer to their question from a series of separate incubators, but decided quite rightly that a TGI would be much more convenient. Nevertheless, it was not until 1960 that the potential merit of a TGI for *bacteriological* studies was recognized (Oppenheimer and Drost-Hansen, 1960). Since this early work many different kinds of organisms have been grown in TGIs including bacteriophage (Landman *et al.*, 1962), actinomycetes (Okami and Sasaki, 1967), yeasts (Battley,

1964), moulds (Fries and Källströmer, 1965; Rowe and Powelson, 1973), algae (Jitts *et al.*, 1964), higher plants (Barbour and Racine, 1967; Fox and Thompson, 1971) and of course bacteria, both aerobic (e.g. Dimmick, 1965; Elliott and Heiniger, 1965) and anaerobic (Cannefax, 1962; Matches and Liston, 1973). However, no review on TGIs appears to have been written and so the purpose of this paper is to bring together the available knowledge on the subject with particular reference to microbes in the hope of encouraging more people to use and develop this elegant methodology.

Theory and Construction of Temperature Gradient Incubators

General principles

The precise construction of a TGI depends both on the organisms which it is intended to accommodate and the characteristics of these organisms it is hoped to determine. Nevertheless, the essential elements of nearly all TGIs are the same, namely a device of uniform cross-section for holding the tubes, petri plates, and other equipment. A heat source is applied to one end only of this device. The only exception to this rule (Okami and Sasaki, 1967) will be described separately. In most TGIs described in the literature the device of uniform cross-section is a metal plate or block, often made of aluminium, on which or into which are placed the organisms under study. Nakae (1966) however, has used a vertical water-filled glass cylinder instead of a metal block, but this seems to lead to unnecessary complications.

When heat is applied to one end of the block a temperature gradient is immediately set up within the block and after a suitable period for equilibration a steady state is achieved whereby the temperature at any point in the block remains constant. The relationship between temperature and distance from the heat source is shown in Fig. 1. Strictly, this relationship is only necessarily true for a solid metal bar, but in practice TGIs exhibit similar properties, i.e. if the metal block is sufficiently well lagged then the temperature of any part of the block is proportional only to its distance from the heat source. Thus it is theoretically only necessary to determine the temperature at two points in the block after equilibrium has been reached in order to predict accurately the temperature at any other point in the block. If the block is not lagged then the relationship between temperature and distance from the heat source is non-linear and follows the general shape given in Fig. 1. An unlagged block is also much more sensitive to changes in ambient temperatures than a lagged block. For these reasons, most TGIs have lagged blocks although some of the

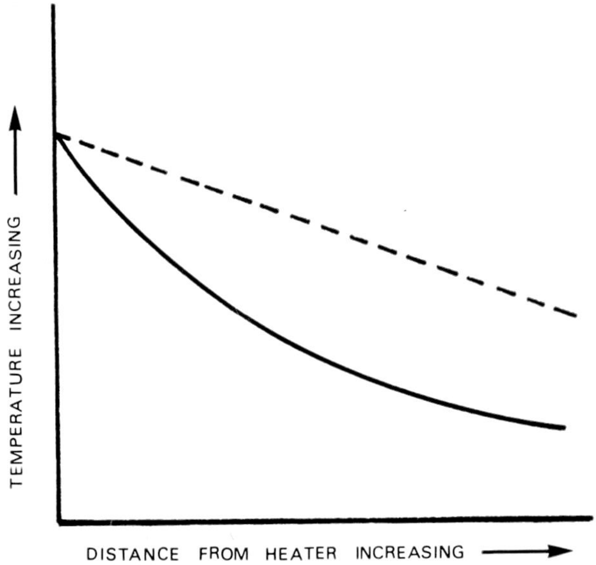

FIG. 1. Typical calibration curves from well-insulated (dashed line) and un-insulated (solid line) temperature gradient blocks.

earlier designs did not include lagging (Cannefax, 1962). No lagging is perfect in so far as some heat always escapes through it. The effectiveness of different sorts of lagging for TGIs has been investigated by Packer *et al.* (1973). The disadvantage of lagging is that the maximum difference in temperature between the two extremes of the block is relatively small unless either the end opposite to the heater is artificially cooled or the block is inordinately long. In practice, therefore, all modern TGIs have some kind of refrigeration applied to that part of the block furthest from the heater. Cooling also enables temperatures below ambient to be maintained.

We now describe a generalized TGI, i.e. an insulated metal block heated to a constant temperature at one end and cooled at the other. There are many variations on this theme, but those TGIs used for studying the activities of microbes may be divided into three groups (a) those in which the temperature gradient is set up along a continuous trough of agar, (b) those in which the microbes are confined in discrete tubes, (c) those in which the microbes are grown in petri dishes.

The agar trough type

In an apparatus of this kind a layer of agar suitable for growing the test organisms is laid along the length of the metal block or plate. The agar can either be placed directly on the sterilized surface of the block (Elliott and Heiniger, 1965) or it can be poured into some removable insert (Battley, 1964; Van Baalen and Edwards, 1973). The apparatus is then allowed to equilibrate before the test organism is inoculated evenly along the entire length of the agar. The advantages of an agar trough TGI are first, it furnishes a continuous temperature gradient so that provided the gradient is stable the organism is subjected to every possible temperature. Secondly, if the agar is covered by a suitable transparent cover the results can be seen at a glance without the need for further measurements such as optical density. The disadvantages are that there are real practical difficulties in operating the apparatus successfully and the results are only of a qualitative nature. The practical difficulties are that sterilization of an entire block with covers can be troublesome, an even inoculation is difficult to achieve and the prepared block acts as a condenser with evaporation occurring at one end and condensation at the other. The evaporation causes the agar to dry out and the condensation reduces visibility through the transparent cover. Battley's (1964) apparatus overcomes the contamination problem by enclosing the agar in sealed tubes.

The first TGI used for microbes (Halldal and French, 1956) was of this type and it was also rather advanced for its time because not only was a temperature gradient employed, but in addition a gradient of light intensity was directed across the agar at right angles to the temperature gradient. Thus a dual gradient was established and the algae growing on the agar were observed for their reactions to two parameters simultaneously. For such dual gradients only one agar surface can be utilized in a single apparatus, but when temperature is the only variable several parallel troughs of agar can be incorporated into the same apparatus. Landman *et al.* (1962) have described just such an apparatus and have also devised ways of dealing with the evaporation and condensation problems.

The discrete tube type

This is the most popular form of TGI and it differs from the agar trough type in so far as the metal-block is drilled to accommodate culture tubes. The tubes may contain agar (Matches and Liston, 1973; Nakae, 1966) or broth (Dimmick, 1965; Packer *et al.*, 1973). The tubes enable the problems of inoculation, contamination and sterilization which bedevil the agar trough TGIs to be eliminated relatively easily. However, the major advantage of tubes containing broth is that growth can be determined quantitatively by absorptiometric or nephelometric methods.

Nevertheless, the use of tubes brings its own problems. Thus, instead of having a continuous temperature gradient the available temperatures are restricted to those of the individual tubes and it is possible that the actual temperature required lies between those of adjacent tubes. Also, if one is working with broths near to the maximum temperature for growth of an organism the growth of a mutant, capable of growing at a temperature higher than the wild type, may not be recognized. The tubes are generally arranged in parallel rows so that either several different organisms can be investigated simultaneously or the study of a single organism can be replicated. A TGI of this type designed by one of the authors (JHB) is shown in Fig. 2.

The discrete tube TGI is ideally suited for studying the growth of anaerobes, but if quantitative growth measurements of aerobic bacteria are to be made then the oxygen must be supplied in sufficient quantity to ensure that temperature, and not the degree of aeration, is the growth-limiting parameter. This means that air must be forced through the broth by some means as the rate of diffusion of oxygen into broth under static conditions is insufficient to maintain maximum growth rates of many bacteria and yeasts. Shaking the tubes individually by hand at regular intervals, as practised by Dimmick (1965), is better than nothing, but is unlikely to be adequate particularly at the higher temperatures. There are two general solutions to the problem of aeration: either the entire gradient incubator must be continuously shaken or air must be pumped through the broth in each tube separately. The incubator of Oppenheimer and Drost-Hansen (1960) and one of those described by Landman *et al.* (1962) were both mounted on shaking machines, but if a simple tube is inserted vertically into the block then a reciprocating action is unfortunately only going to aerate adequately the upper part of the broth. For example, at a shaking speed of 1·4 c s^{-1} a drop of crystal violet solution in 10 ml distilled water in a 13 mm diam. test tube was not adequately mixed after 15 min of continuous shaking.

Landman *et al.* (1962) improved the degree of aeration created with a reciprocal shaking motion by inserting the tubes at an angle of 45° into the block. However, because the temperature of the tube is a function of its distance from the heat source these workers have introduced a much greater variation of temperature within a tube than occurs when the tube is inserted vertically.

Aeration by means of a bubbled air supply has been used by both Palumbo *et al.* (1967) and Baker (1974). The two systems have many features in common; in both the air supply is split into as many supplies as there are rows of tubes in the block and the air is first humidified and brought to the temperature of the particular broth to be aerated before it

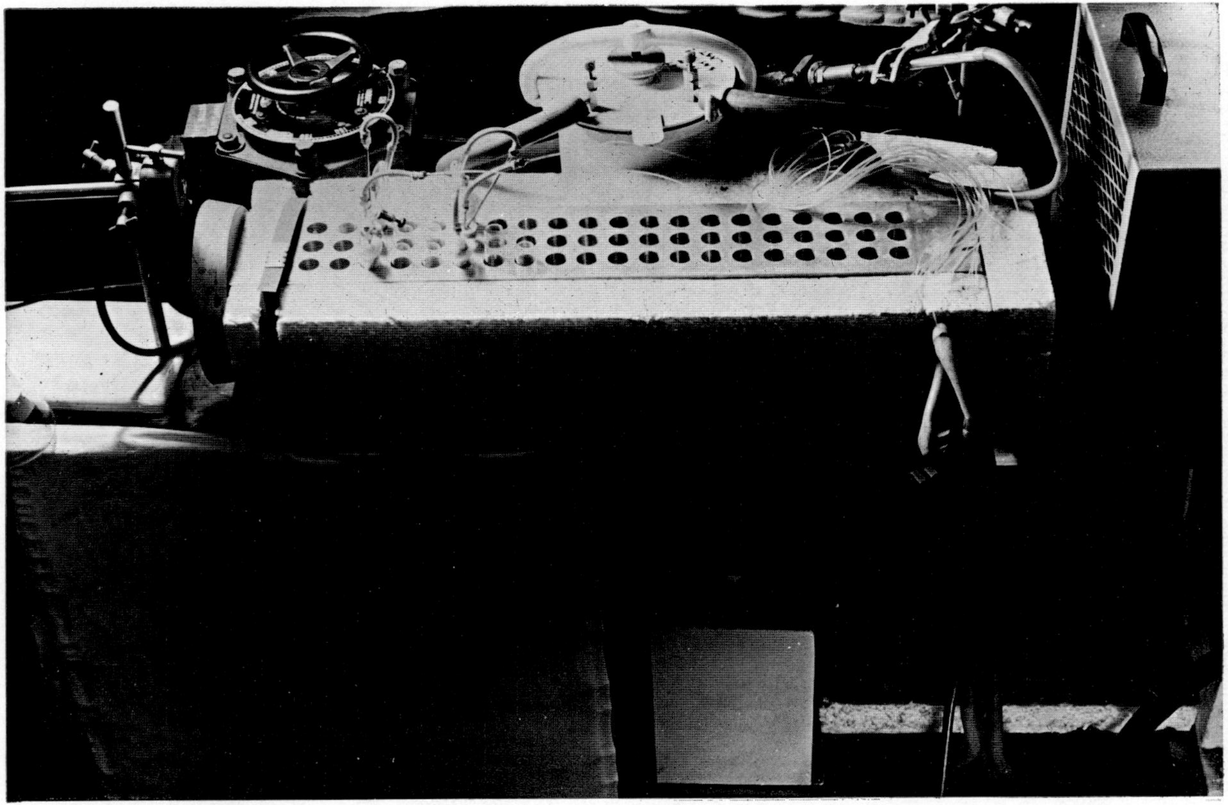

FIG 2. A temperature gradient incubator of the discrete tube type. The upper insulation and many of the tubes, have been removed to reveal the aluminium block. The wheel on the upper left regulates the voltage to the hotplate. Refrigerant is pumped from the bath below the block and the membrane filter holder held by the clamp leads to the rubber bulb manifold.

is introduced. The latter procedure is important because dry air would markedly increase evaporation and the air supply must not affect the temperature of the broth. In the apparatus of Palumbo *et al.* (1967) the air is filtered by a cotton wool plug in each tube, but in Baker's (1974) apparatus the air is filtered through a single membrane (0·2 μm pore size) before it reaches the manifold. The advantage of the former apparatus is that there is less likelihood of contamination of the broth when it is removed from the block for a cell density determination, but Baker (1974) did not find this to be a problem. On the other hand, membrane filters are a much more efficient way of removing airborne microbes than cotton wool and the aeration supply is easier to set up.

Petri dish type

Petri dishes can be placed on any temperature gradient block and provided they are adequately insulated from the inconsistencies of the laboratory environment they may serve their purpose very well. TGIs of this sort have been described by Smith and Reiter (1974) and Van Baalen and Edwards (1973); however, the standard petri dish is 9 cm in diam. and so the dish will itself have a temperature gradient across it. Thus, no petri dish can be taken as representative of a single given temperature. Fluegel (1963) overcame this problem very neatly by making a vertical temperature gradient block as opposed to a horizontal one. The petri dishes were placed in slots cut in the block so that instead of having a temperature gradient across their diameter it occurred across their depth which is no worse than that across a tube.

Heating and cooling

Methods for heating one end of a TGI can be divided into two groups: (a) those in which some form of electrical heater is directly applied to the block, (b) those in which heat is supplied from a fluid reservoir.

Electrical heaters which have been used include soldering iron elements (Dimmick, 1965), heating tape (Smith and Reiter, 1974) and a hotplate (Baker, 1974). Fluid reservoirs have been used by many workers. Cannefax (1962) supported his block with a metal leg at each end and one leg was placed in a heated oil bath; however, most TGIs using fluid reservoirs have the fluid (generally hot water) pumped through holes drilled in the end of the block.

Ideally, whichever heating system is used two conditions should be met. First, there should be no detectable temperature variation within any plane in the block parallel to the heated end and secondly, there should be no detectable change in temperature with time at any given point in the block after equilibrium has been reached. No practical instrument can comply with these conditions absolutely, but some methods of heating are theoretically more likely to give a closer approximation than others. Thus to establish a uniform temperature in a cross-section the entire face of the block needs to be heated to the same temperature. Of the above types of heating only the hotplate is capable of doing this although in practice an electrical hotplate does not have a uniform surface temperature itself and so other methods may be equally good. Nevertheless, the nearer this condition is approached the better, so that several soldering iron elements should be used instead of one, and as many holes as possible should be drilled (in the same plane) if pumped water is used. Moreover, to minimize temperature variations the pumping rate should be increased to the maximum available.

In order to satisfy the second condition, i.e. maintaining a constant temperature with time, some thermostatic device is usually included in the heater circuit. Unfortunately most thermostats are relatively simple so-called "on-off" devices by which it is meant that the heater is on maximum power or not on at all. Consequently, when the required temperature is reached and the thermostat switches the heater off there is a considerable quantity of residual heat left in the heater which gives rise to an "overshoot" of the required temperature. Thus the temperature fluctuates in a characteristic saw-tooth fashion as shown diagrammatically in Fig. 3(A). For precise temperature control the amplitude of the saw-tooth must be minimized and there are two different ways of doing this. The first method is to control the heat input not by means of a thermostat, but by reducing the voltage across the heater until the heat input exactly matches the heat loss. At this point the system is perfectly balanced and because the heat input is continuous there is no saw-tooth effect. The disadvantages of this form of temperature control are that the TGI needs to be in a controlled temperature room and that very long periods may be required before the balance point is reached at any given voltage. Nevertheless, it is possible to combine voltage control with an on-off thermostat (Baker, 1974) and in such cases the voltage is kept slightly higher than the balance point and the amplitude of the saw-tooth is greatly reduced.

The second method of precise temperature control employs a different kind of electrical thermostat called a proportional device. In this system shown in Fig. 2(b) the power in the heater is progressively reduced as the required temperature is approached. Thus, the overshoot is very much less than with the on-off type of thermostat. Proportional temperature controllers have only recently become generally available, but their

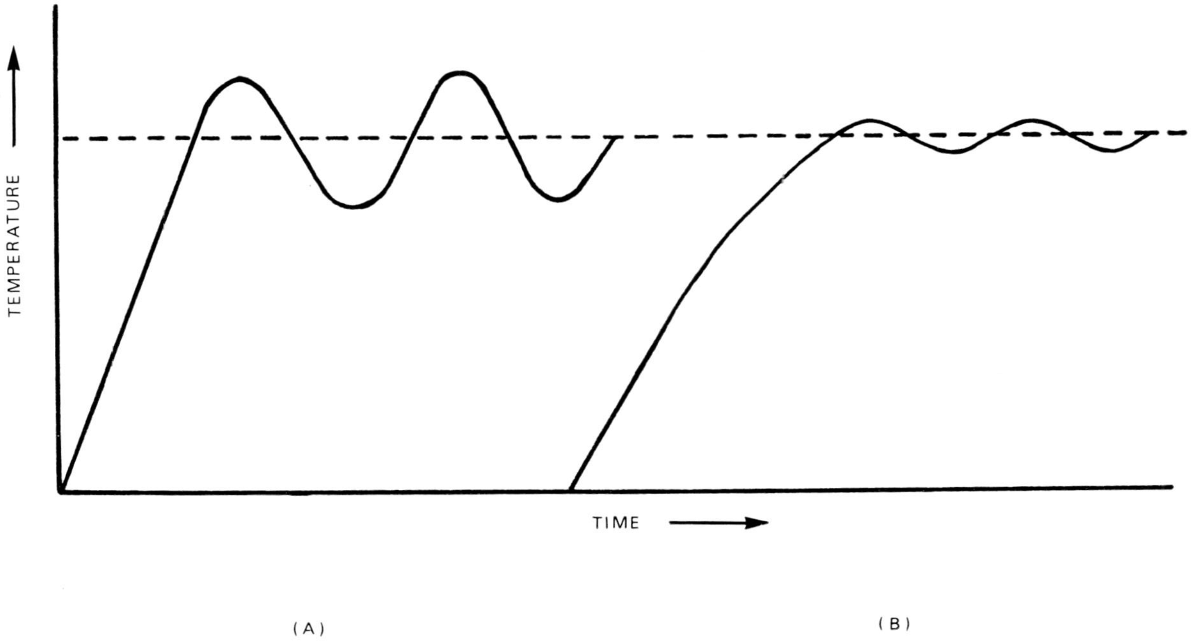

FIG. 3. Response curves of two different types of thermostat. The on-off device (A) results in a linear approach to the desired temperature (dashed line) and a subsequent broad oscillation. In comparison the proportional device (B) causes the temperature to rise progressively more slowly with a relatively small subsequent oscillation.

advantages suggest that they will soon become commonly used.

Cooling systems for the opposite end of the gradient block to the heater are governed by much the same conditions as the heating systems. Some of the earlier designs (Landman *et al.*, 1962) employed no special cooling system, but as stated earlier such a system is only practicable with an uninsulated block and then the temperature gradient is non-linear. Dimmick (1965) describes a somewhat complicated cooling system employing 2-methoxyethanol and dry ice, but most TGIs have a refrigerant pumped through holes drilled in the end of the block. Fluegel (1963) used tap water instead of a special pumped supply, but most water supplies vary rather more in temperature than Fluegel's did. The temperature of the cooling system should be controlled in a precise way similar to the heating system.

Okami and Sasaki (1967) designed a novel TGI in which the temperature gradient was established not according to rates of cooling, but according to the relative quantities of hot and cold water making contact with the tube. Their apparatus consisted of a rectangular box divided across one diagonal. Culture tubes were inserted through the side of the box such that part was on one side of the diagonal divider and part on the other. Hot water was pumped through one end of the box and cold water through the other end. When the entire apparatus was shaken with a wrist action on an axis perpendicular to the culture tubes the temperature of an individual tube depended on how much of it was exposed to the hot, and how much to the cold, water. Nevertheless, this ingenious device does not seem to have any intrinsic advantages over its rivals and has not been developed further.

Temperature determination and fluctuation

The temperature at any point in a TGI can be determined by any type of thermometer. Mercury-in-glass thermometers, which are the simplest to use, have been employed by Baker (1974) and Nakae (1966). Thermocouples (Fluegel, 1963; Van Baalen and Edwards, 1973) and thermistors (e.g. Cannefax 1962; Okami and Sasaki, 1967) have the advantage that they can be used to record temperatures automatically. It is important to determine the temperature at which the organism is growing rather than the temperature of the adjacent block. Thus the temperature of broth in a tube may be slightly different from the temperature of the block which surrounds it. To minimize this difference, water, oil or mercury may be usefully placed in the bottom of each hole into which the tube is later placed. If mercury is used high temperatures should not, of course, be reached at any time.

Ordinary bacteriological incubators commonly fluctuate in temperature by ± 2°C at any given setting with time. One of the big advantages of a TGI is that temperature fluctuation with time is normally much less than this although such precision is not always necessary. Two types of temperature change should be distinguished: that between replicates at any given time (known as temperature *variation*) and that between different times at any given point (known as temperature *fluctuation*). Temperature variation, for example between replicate rows of tubes, will always occur due to the different distances of the replicate tubes from the sides of the TGI. Nevertheless, careful design can reduce this change to tolerable limits and Packer *et al.* (1973) have reduced it to 0·07°C. Temperature fluctuation is not inherently necessary unlike temperature variation and although it is sometimes measurable a good apparatus will keep it to < 0·1°C. Illuminated TGIs are more liable to unintentional temperature changes. Jitts *et al.* (1964) like Halldal and French (1956) used water as a heat filter in front of these lamps. Both temperature variation and temperature fluctuation should be carefully recorded before any TGI is used for experimental purposes. Unacceptable values for either can then be eliminated or allowed for. Placing the TGI in a temperature controlled room generally improves its performance even when the amount of insulation seems adequate. The most satisfactory way of conducting an experiment is to determine temperatures at different points in the incubator several times during the actual course of the experiment. This can be done by sacrificing one row of culture tubes and putting some type of thermometer in each. If for some reason this is impossible then the temperature gradient should be at least checked before and after the experiment.

Commercially Available Temperature Gradient Incubators

There are at least two different TGIs made commercially which are available in the UK. The Toyo Kagaku Sangyo Co Ltd, Tokyo, makes the TN-3 model supplied by ChemLab Instruments Ltd, Hornchurch, and Scientific Industries Inc, New York, with a UK subsidiary in Loughborough, markets another model. Both incubators are of the discrete tube type although they can be simply modified to take agar troughs. Both provide tubes at 30 different temperatures and tackle the aeration problem by shaking the block with a wrist-action. The shaking action is made much more efficient by the use of L-shaped tubes because the liquid culture is in the horizontal part of the tube. Tubes of this type were also used by Okami and Sasaki (1967) and deserve to be used more widely. Commercial TGIs are, of course, convenient, but they suffer from the disadvantages that the TN-3 does not apparently cater for replicates at all and the other model can only accommodate two tubes at any given temperature, also in February 1978 they each cost approximately £3000. A moderate laboratory workshop can produce a TGI to match one's requirements very much more cheaply.

Applications of Temperature Gradient Incubators

One of the most appealing characteristics of TGIs is their great versatility; they may be used over the entire range of growth temperatures from sub-zero for psychrophiles to the maximum growth temperature of thermophiles. Depending on the organism and growth medium results may be recorded in terms of dry weight produced, optical density or colony counts. Up to the present the most frequent application of TGIs has been to determine cardinal temperatures for growth of microbes. Maximum and minimum growth temperatures may be determined from either liquid or agar cultures, but optimal growth temperatures (or temperature ranges) may only be satisfactorily determined using liquid cultures. Some workers (Oppenheimer and Drost-Hansen, 1960) claim that multiple temperature optima may occur under certain conditions. The extremes of growth are often not clearly defined on agar and Battley (1964) has described in detail an effective method of dealing with this problem. In liquid media the growth of a mutant may interfere with the determination of minimum and maximum growth temperatures of a specific organism. Maximum growth temperatures can also depend on the composition of the medium (Fries and Källströmer, 1965).

The optimum temperature for growth is not necessarily identical with the optimal temperature for other physiological functions. Thus, Nakae (1966) showed, with the aid of a TGI, that optimum acid production by *Streptococcus lactis* occurred at a different temperature to optimal growth. Similarly, optimal antibiotic production by a *Bacillus* happens at a sub-optimal growth temperature (Okami and Sasaki, 1967). Smith and Reiter (1974) used a TGI to determine the optimum temperature for sporulation of some fungi and many other physiological properties could be investigated.

Besides cardinal temperatures, a TGI is very useful for the determination of growth rates at many different temperatures. Liquid cultures must be used for quantitative work and aerobic cultures must be adequately aerated. Growth rates are calculated from nephelometric or absorptiometric measurements and so it is convenient to make the culture tubes in the TGI compatible with the nephelometer or other optical system. Otherwise, subsamples may have to be removed thereby increasing the risk of contamination and increasing the total volume of each culture required. In addition to growth rates, death rates can also be important, e.g. in food preservation studies, and Elliott and Heiniger (1965) have used a TGI at supra-maximal growth temperature to determine differences in death rates of various salmonellae.

Quite a different application of a TGI concerns the isolation of temperature insensitive mutants, i.e. strains which grow at temperatures above the maximum of the wild type. Landman *et al.* (1962) obtained mutants of *Bacillus* in this way capable of growth at 2–3°C above the maximum of the wild type, but were unable to obtain a true thermophile from a mesophilic wild type. The least used application of TGIs, and one with considerable potential, is a combination of a temperature gradient with a different crossed gradient. To date, only algologists seem to have made use of this idea with crossed gradients of temperature and light (Van Baalen and Edwards, 1973; Jitts *et al.*, 1964). The former authors consider the method to have a predictive potential with regard to algal blooms, but there are many other possible cross-gradients. For example, a salinity cross-gradient could be used for estuarine microbes and antibiotic cross-gradients could also yield worthwhile results.

In conclusion, we have shown that temperature gradient incubators are valuable research tools. They are not difficult to construct provided their future use is borne in mind at the design stage, or they can be obtained commercially. A TGI is a versatile apparatus that may be used for biochemical, physiological and taxonomic studies on a wide range of organisms. It saves utilizing many standard incubators and often enables experiments to be carried out which would otherwise be impossible.

Acknowledgement

We are grateful to G. I. Williams for helpful discussion.

References

Baker, J. H. (1974). The use of a temperature gradient incubator to investigate the temperature characteristics of some bacteria from Antarctic peat. *British Antarctic Survey Bulletin* **39,** 49–59.

Barbour, M. G. & Racine, C. H. (1967). Construction and performance of a temperature-gradient bar and chamber. *Ecology* **48,** 861–863.

Battley, E. H. (1964). A thermal-gradient block for the determination of temperature relationships in microorganisms. *Antonie van Leeuwenhock* **30,** 81–96.

Cannefax, G. R. (1962). A temperature-gradient bar and its application to the study of temperature effects on the growth of Reiter's treponeme. *Journal of Bacteriology* **83,** 708–710.

DIMMICK, R. L. (1965). Injury and growth of *Serratia marcescens* studied in a thermal gradient incubator. *Applied Microbiology* **13,** 846–850.

ELLIOTT, R. P. & HEINIGER, P. K. (1965). Improved temperature gradient incubator and the maximal growth temperature and heat resistance of *Salmonella. Applied Microbiology* **13,** 73–76.

FLUEGEL, W. (1963). A shelf-type gradient incubator. *Canadian Journal of Microbiology* **9,** 859–862.

FOX, D. J. C. & THOMPSON, P. A. (1971). A thermo-gradient apparatus designed for biological investigations over controlled temperature ranges. *Journal of Experimental Botany* **22,** 741–748.

FRIES, N. & KÄLLSTRÖMER, L. (1965). A requirement for biotin in *Aspergillus niger* when grown on a rhamnose medium at high temperature. *Physiologia plantarum* **18,** 191–200.

HALLDAL, P. & FRENCH, C. S. (1956). The growth of algae in crossed gradients of light intensity and temperature. *Carnegie Institute Washington Year Book* **55,** 261–265.

HERTER, K. (1934). Eine verbesserte Temperaturogel und ihre Anwendung auf Insekten und Saugetiere. *Biologisches Zentralblatt* **54,** 487–507.

JITTS, H. R., MCALLISTER, C. D., STEPHENS, K. & STRICKLAND, J. D. H. (1964). The cell division rates of some marine phytoplankters as a function of light and temperature. *Journal of the Fisheries Research Board of Canada* **21,** 139–157.

LANDMAN, O. E., BAUSUM, H. T. & MATNEY, T. S. (1962). Temperature gradient plates for growth of micro-organisms. *Journal of Bacteriology* **83,** 463–469.

MATCHES, J. R. & LISTON, J. (1973). Temperature-gradient incubator for the growth of clostridia. *Canadian Journal of Microbiology* **19,** 1161–1165.

NAKAE, T. (1966). Method of temperature-gradient incubation and its application to microbiological examinations. *Journal of Bacteriology* **91,** 1730–1735.

OKAMI, Y. & SASAKI, Y. (1967). Temperature-gradient tools. *Applied Microbiology* **15,** 1252–1255.

OPPENHEIMER, C. H. & DROST-HANSEN, W. (1960). A relationship between multiple temperature optima for biological systems and the properties of water. *Journal of Bacteriology* **80,** 21–24.

PACKER, G. J. K., PRENTICE, G. A. & CLEGG, L. F. L. (1973). Design of a temperature-gradient incubator. *Journal of Applied Bacteriology* **36,** 173–177.

PALUMBO, S. A., BERRY, J. M. & WITTER, L. D. (1967). Culture aeration in a polythermostat. *Applied Microbiology* **15,** 114–116.

ROWE, R. C. & POWELSON, R. L. (1973). A temperature-gradient plate designed to function near and below 0°C. *Phytopathology* **63,** 287–288

SELWYN, M. J. (1961). An apparatus for maintaining a range of constant temperatures. *Biochemical Journal* **79,** 38.

SMITH, J. D. & REITER, W. W. (1974). A general-purpose illuminated temperature-gradient plate. *Canadian Journal of Plant Science* **54,** 859–864.

VAN BAALEN, C. & EDWARDS, P. (1973). Light-temperature plate. In *Handbook of phycological methods* (Stein, J. R., ed.). Cambridge: Cambridge University Press, pp. 267–273.

Determination of Heat Resistance of Cold Tolerant Sporeformers by means of the "Screw-cap Tube" Technique

M. J. M. Michels

Department of Microbiology, Unilever Research Duiven, Zevenaar, The Netherlands

Introduction

Cold tolerant sporeforming bacteria, which are known to occur in the environment (Sinclair and Stokes, 1964; Larkin and Stokes, 1966; Michels and Visser, 1976) are important as potential spoilage organisms for pasteurized foods that rely on refrigeration for shelf life. The best known example of such food is non-reinfected pasteurized milk, of which the shelf life of two weeks or more at commercial refrigeration temperatures (5–7°C) is determined by cold tolerant sporeformers, although some non-sporing thermoduric organisms may also be of significance (Mourgues and Auclair, 1973; Washam *et al.*, 1977). These cold tolerant sporeformers from milk (mainly *Bacillus* sp.) have been studied fairly well (Grosskopf and Harper, 1969; Shehata and Collins, 1971; Shehata *et al.*, 1971; Bhadsavle *et al.*, 1972; Langeveld *et al.*, 1973; Grosskopf and Harper, 1974) and the heat resistance of the spores of these bacteria has been considered to be an important characteristic (Shehata and Collins, 1972; Bhadsavle *et al.*, 1972; Washam *et al.*, 1977).

Another, lesser-known, type of pasteurized food, the shelf life of which depends on refrigeration, consists of meals for institutional catering, preserved according to the Nacka process, which received its name from the Swedish hospital where this process was first applied (Delphin, 1971). Such ready-to-eat meals, which according to a typical process are packed in plastic pouches, pasteurized for 5–10 min at 80°C and cooled in ice-water, have a shelf life at 0–3°C of about 3 weeks (Tändler, 1972). For the study of the heat resistance of cold tolerant *Bacillus* sp., which determine the shelf life of such Nacka meals, we found the "screw-cap tube" technique most suitable. This technique was developed in our laboratory as a simple, convenient and accurate method for the determination of the thermoresistance of bacterial spores (Kooiman, 1974; Kooiman and Geers, 1975). We shall describe this technique in its original simplest form together with some modifications which allow for an improved ease of operation and a wider field of application.

This screw-cap tube technique is suitable only for the determination of the heat resistance of micro-organisms in a liquid environment; those interested in the dry heat resistance of cold tolerant sporeformers are referred to the paper of Winans *et al.* (1977).

Production of Spore Suspensions for Heat-resistance Studies

Several sporulation media have been described for the production of spores of cold tolerant sporeformers. Table 1 lists media and incubation conditions used for growing spore crops for heat-resistance studies. The nutrient agar of Shehata and Collins (1972) gave better spore yields than eight other media which they tested and this medium is therefore likely to be a suitable choice, although a particular group of the cultures used by Shehata and Collins did not sporulate well on this medium or any of the other media tested. The incubation temperature for spore production is usually chosen as that which is close to the optimum temperature of the strain under study; for most strains this is *ca.* 20°C (Larkin and Stokes, 1966; Laine, 1970). Nevertheless, lower temperatures can be selected, if required, because spore formation can be observed for most strains at temperatures close to their minimum temperature for growth.

In general the following procedure is followed for the production of a batch of spores. The surface of agar plates or slants of a suitable sporulation medium is inoculated with sufficient cells (or preferably with a heat-shocked spore suspension) to give heavy growth on the agar under the incubation conditions chosen. During incubation, the culture is checked regularly under the phase-contrast microscope for the presence of spores. When most cells have sporulated (preferably over 90%), the spore crop is harvested from the agar with cold, sterile, distilled water and the spore crop is washed at least three times by alternatively centrifuging and resuspending the cells. The resulting spore crop is finally resuspended in a small volume of distilled water and usually stored at a temperature of 2–4°C.

For cold tolerant sporeformers we prefer, as an alternative to refrigeration, storage of the spore crop in 2 ml portions at −25°C, to make sure that no change in heat resistance occurs at prolonged refrigerated storage. For heat-resistance studies, a 2 ml portion containing *ca.* 10^8

TABLE 1. Conditions for production of spores of cold tolerant Bacilli

Sporulation medium	Incubation Temperature (°C)	Incubation Time (d)	Reference
Potato Infusion Agar + 7 inorganic salts (used as agar slants)	20 (1·1)[a]	10 (70)	Marshall and Ohye (1966)
Trypticase Soy Agar (Oxoid) + 10 mg $MnSO_4$ l^{-1} (used as agar slants in Roux-bottles)	20 (2)		Laine (1970)
Nutrient Agar (BBL) + 0·05% $MgSO_4$ and 0·05% $MnSO_4$ (used as agar slants in bottles)	25–30	7–10	Shehata and Collins (1972)
Trypticase Soy Agar (BBL)	20 (5)	5–8 (14–35)	Michels and Visser (1976)
AK-2 Agar (BBL) + 0·8 ml of 10% $CaCl_2$ 100 ml^{-1}	7	3–7	Winans *et al.* (1977)

[a] Numbers in parentheses list conditions of temperature (°C) or time (d) for spore production at low temperatures.

spores ml^{-1} is thawed and is then ready for use. The remaining part of the thawed portion is kept at 0–4°C and used up within a week.

The Original Screw-cap Tube Technique

In the screw-cap tube technique, spores are heated to a given temperature by injection of 0·1 ml of a concentrated spore suspension into the contents of a Pyrex glass tube through a septum fitted in the screw cap. The tube's contents consist of 10 ml of a suitable heating menstruum, e.g. sterile phosphate buffer, preheated to the selected exposure temperature in a thermostatic bath. At the end of the selected heating time the Pyrex tube is transferred from the heating bath and cooled in ice-water.

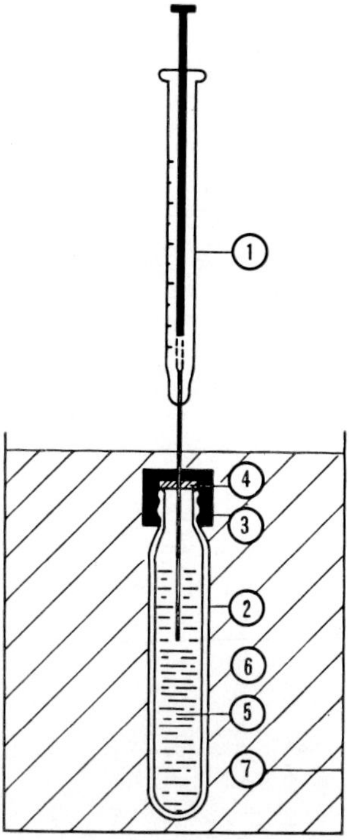

FIG. 1. Design of the original screw-cap tube technique: 1, Hamilton, microlitre syringe, Model 710N, capacity 100 μl; 2, glass tube, Pyrex, No. 9825, 16 × 100 mm; 3, screw-cap, provided with a central hole of 2 mm diam.; 4, septum, 13 mm diam., 3 mm thick (e.g. Hewlett Packard "low bleed" neoprene, 5080-6722); 5, heating menstruum (e.g. sterile phosphate buffer); 6, heating liquid (water or glycerol); 7, thermostatic bath (e.g. Tamson, Zoetermeer, The Netherlands. TXB9-150).

With this technique the heating up of the spores to the exposure temperature is almost instantaneous, and there is also a very rapid cooling at the end of the exposure period. (The deviations from the ideal heating and cooling curve which still occur will be considered in a following section.) The design of the system is given in Fig. 1. The tube consists of a standard Pyrex glass tube with a screw-cap, which has been modified by having a small hole drilled in the centre. A heat stable septum within the cap allows for injection of spore suspension without loss of internal pressure in the tube. In a typical heat resistance study 6 previously sterilized screw-cap tubes, filled with 10 ml of phosphate buffer, are placed in a rack in a thermostatic bath at a suitable temperature (e.g. 90° for cold tolerant sporeformers). These tubes should be completely submerged and kept at bath temperature for 15 min to heat the contents to this temperature. The first injection of 0·1 ml of spore suspension was made into tube No. 6 immediately and successive injections are made after 5, 10, 15, 20 and 25 min in tubes No. 5, 4, 3, 2 and 1, respectively. After 30 min, the rack with the tubes is taken from the heating bath and cooled in ice-water (preferably sterile ice-water should be used to prevent accidental contamination of the contents when the tubes are opened). The spores in tubes 1 to 6 have now been exposed to 90°C for 5, 10, 15, 20, 25 and 30 min respectively. The number of surviving spores per tube is determined with a suitable recovery medium and these data can be used to calculate the D-value of the spores under study.

Modifications to the Screw-cap Tube Technique

In two years of experience with the screw-cap tube technique of Kooiman (1974), we developed three modifications to the original technique.

1. A Hamilton spring-driven push-button syringe, model CR 700–200 replaced the standard Hamilton syringe, model 710 N. With this type of syringe, the possibility of the plunger being pushed up by the pressure in the tube—after the needle has passed through the septum—is prevented. This change simplified the procedure and was less elaborate than the modification described by Hauschild and Hilsheimer (1977).
2. Following the work of Kooiman (1974), we have developed stainless steel screw-cap tubes to be used for heat-resistance studies at exposure temperatures well over 100°C, as there is always a minute chance of a glass tube bursting by the internal pressure generated in the tube by the vapour pressure of water. That this risk with undamaged tubes is likely to be small was demonstrated by a test in which glass screw-cap tubes were subjected to an effective internal pressure of 600 kPa,

without any damage to the tubes, whereas the effective vapour pressure generated by water at 120°C is not more than 100 kPa. Notwithstanding this observation, we felt that at exposure temperatures over 120°C a steel tube is to be preferred, as such a tube is safe under all operational conditions.

3. The third modification was the introduction into the tubes of a small teflon-coated magnetic stirring rod (Cenco Stirring Bar, 18854–11, 22 × 8 mm). Positioning of the screw-cap tubes, provided with such a rod, in a circular rack around a permanent magnet (Eclipse, Neill and Co, Sheffield, Cat. No. 812 B) which is rotated by a laboratory stirrer at 300 r min^{-1}, allowed for an excellent homogenization of the contents of the tubes during the entire exposure period. With this modification the time required for the tubes' contents to recover from the small temperature drop caused by injection of the cold spore suspension was reduced considerably (for details see next section). The rapid homogenization of the contents of the tubes and the improved thermal recovery to exposure temperature have extended the application of this technique to exposure times of 30 s and less. Another potential of this modification is its application for heat-resistance studies of spores in slightly viscous heating menstrua.

The design of the screw-cap tube system with these three modifications is shown in Fig. 2. The propellor blade below the Eclipse magnet is used to improve the flow of the heating liquid around the tubes. The circular rack which contains the tubes has a capacity of 8 tubes, positioned on a circle with a diam. of 67 mm, this number being amply sufficient for the determination of a spore survivor curve.

Temperature Profile for Injected Spores

The injection of 0·1 ml of spore suspension at 0°C into a screw-cap tube containing 10 ml of phosphate buffer at e.g. 100°C causes an immediate rise of the temperature of the spores to almost 100°C. A temperature identical to the exposure temperature is not directly realized because the injection of the cold spore suspension causes temporarily a small drop in temperature, which can be calculated to be 1°C for 10 ml of buffer at 100°C (Kooiman and Geers, 1975).

With thermocouples placed at the centre of a screw-cap tube we have measured the effect of our modifications on the temperature drop caused by injection of the spores in the tubes. These measurements have been made for an exposure temperature of 90°C only, as this is a suitable temperature for heat-resistance studies with cold tolerant sporeformers and moreover the calculated temperature drop for spores heated at 80–

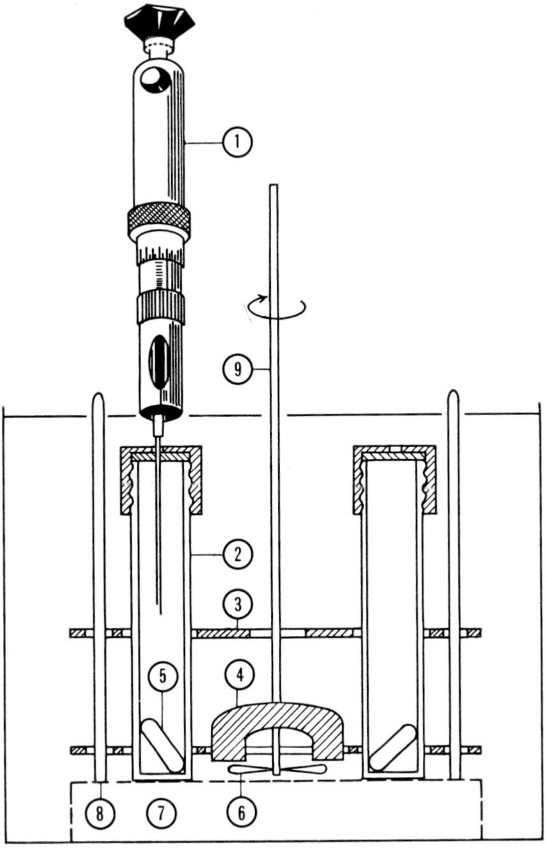

FIG. 2. Design of the modified screw-cap tube technique: 1, Hamilton push button adjustable syringe, model CR 700-200; 2, stainless steel tube, 16 × 107 mm, with a screw-cap with central hole and heat-stable septum; 3, circular rack for placing 8 tubes around magnet; 4, permanent magnet, Eclipse, Cat. No. 812 B (Neill and Co, Sheffield); 5, magnetic stirring bar, Cenco, No. 18854-11, 20 × 8 mm; 6, propellor blade; 7, support frame for rack; 8, guiding rod for positioning of rack in heating bath; 9, magnet shaft to be rotated via laboratory stirrer at 300 r min^{-1}.

120°C differs only slightly, i.e. 0·8–1·2°C. The injection of 0·1 ml of spore suspension at 0°C into 10 ml of phosphate buffer at 90°C was done with the Hamilton CR 700–200 syringe and the temperature drop was recorded for a glass and a steel screw-cap tube, both with and without the described stirring device. After a heating period of 5 min the tubes were taken from the heating bath and cooled immediately under shaking in ice-water.

The results of these measurements with a steel screw-cap tube are represented in Fig. 3 as a temperature profile for the spores. Measurements with the glass tubes were almost identical to those for the metal tubes. Figure 3 shows that with the stirring device the time for the contents of the tubes to recover completely from the initial temperature drop was reduced by more than 50%, i.e. from 2·5 min without stirring to about 1 min with stirring. Another point which is made clear in Fig. 3 is that at the end of the heating period the cooling of the spores is very rapid; in fact the temperature dropped 10°C in about 2 s.

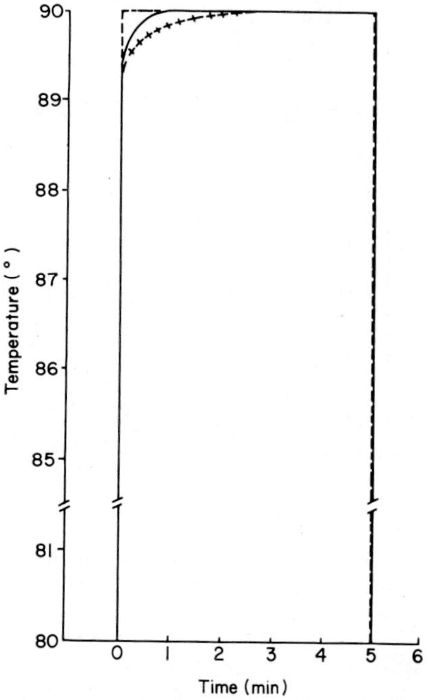

FIG. 3. Temperature profile for the upper 10°C part of the heating curve of 0·1 ml of spore suspension injected into 10 ml of phosphate buffer at 90°C in a steel screw-cap tube, with (————) and without (+++++) a stirring device (300 r min^{-1}).

For both exposure systems, i.e. with and without the stirring device, we have calculated which adjustment of the heating time would be required to correct for the small temperature lag met by the spores at the start of the heating period. For this purpose, the temperature profile of Fig. 3 was plotted on specially ruled lethal-rate paper with a temperature scale of 70–90°C (Stumbo, 1965) and from the resulting graph it was deduced that for spores heated without stirring at 90°C each heating time should be reduced by 6 s to give the time at 90°C which was effectively met by the spores. With the stirring device this correction was reduced to 2·5 s. If we take into account also the time lag met by the spores at the start of the cooling, which corresponds effectively to about +1·5 s, then the total correction for the stirred screw-cap tube technique is 1 s (−2·5 and +1·5 s).

In the example given, the correction was calculated for spores injected into phosphate buffer at 90°C and because the temperature drop at other heating temperatures is only slightly different (0·8°C at 80°C and 1·2°C at 120°C) the corresponding correction times for the stirred screw-cap tube technique will be very close to the already mentioned one second. In other words it can be stated that with the stirring device there will be no need to make a correction for the heating up and cooling down lags unless very short heating times are chosen, such as $<$ 30 s.

Enumeration of Spores and Determination of D-values

In contrast to heat-damaged spores of cold tolerant *Clostridium botulinum*, which require lysozyme for an optimal recovery (Alderton *et al.*, 1974), there are no such requirements known for cold tolerant food spoilage sporeformers for which, therefore, general purpose media can be used. Shehata and Collins (1972) enumerated surviving spores with Difco Standard Plate Count Agar plus 0·1% starch and an incubation of 2 d at 30°C. Another suitable medium is formed by Trypticase Soy Agar with incubation of the plates for 3–5 d at 20 or 21°C (Michels and Visser, 1976; Winans *et al.*, 1977). For the recovery of spores of a cold tolerant strain of *Cl. hastiforme*, thioglycolate agar with 0·1% starch and an anaerobic incubation for 4 d at 25°C was used by Bhadsavle *et al.* (1972). Also, incubation temperatures below the optimum temperature for growth have been used and although sometimes no difference in survivor curves was found with incubation at 20 and 1·1°C (Marshall and Ohye, 1966) there certainly are strains for which recovery at 0 or 5°C results in much lower survivor counts (Michels and Visser, 1976).

From the spore counts a survivor curve can be constructed by plotting the log of the survivor counts versus the corresponding heating times (Stumbo, 1965). From the straight part of this curve the D-value of the spores under study can be read easily, but this method does not give very accurate D-values as such a survivor curve is drawn merely as the apparently best fitting line to the survivor counts. A better practice is to calculate this line with a least square regression analysis, which can be performed easily since the introduction of the electronic calculators of which, e.g., a Texas Instruments SR-51A contains a suitable programme. Preferably also the 95% confidence limits to the D-value should be calculated, but this requires more statistical skill or a suitable computer programme. An example of a study in which D-values with 95% confidence limits were calculated is formed by that of Winans *et al.* (1977). These authors prefer to calculate the D-value without using N_0 (the number of spores at the start of the heating period) because in this way an initial shoulder or drop in the spore survivor curve no longer affects the resulting D-value. This practice could be interesting for those workers who might see a problem in the small correction for temperature lag which could be required in very accurate studies with the original screw-cap tube technique. If N_0 is not taken into account and the required correction of the heating time is the same for all heating times, then the D-values based on actual heating times and on corrected heating times will be the same, which in other words eliminates any need for the use of such a correction.

Discussion

As shown in Table 2, a great deal of information exists on the heat, resistance of cold tolerant sporeformers, which could be helpful for those workers planning further research in this area. One should, however, not conclude too much from the available data, which give D_{90° values for cold tolerant spores of approximately 1–10 min, as all the strains mentioned in this table were isolated after pasteurization of a food or environmental sample for 5 or 10 min at 80°C (Larkin and Stokes, 1966; Marshall and Ohye, 1966; Shehata and Collins, 1971; Bhadsavle *et al.*, 1972; Michels and Visser, 1976) and these isolation conditions could have eliminated the more heat-sensitive spores of cold tolerant sporeformers. That less heat-resistant strains can occur is shown by the cold tolerant *Cl. botulinum* type E with a D_{80° of about 1 min, whereas Washam *et al.* (1977) isolated cold tolerant sporeformers from milk, of which the spores survived HTST pasteurization conditions (such as 16 s at 71·7°C), but which were inactivated by pasteurization for 30 min at 60·7°C. Using a z-value of 10°C, this corresponds with inactivation after heating for 3 min at 70°C or 0·3 min at 80°C and such strains would therefore have been missed in the previously mentioned studies.

For heat-resistance studies with cold tolerant sporeformers the screw-cap tube technique is most conveniently used in the form of a glass screw-cap tube with a Hamilton push-button syringe. In this form the system was found to be very suitable also for studies on other bacterial

TABLE 2. Heat resistance data for cold tolerant sporeformers[a]

	$D_{90°}$ (min)	z-value (°C)		$D_{90°}$ (min)	z-value (°C)
B. circulans	4·4	—	*Bacillus* Cp-5	1·6	6·6
B. brevis	4·8	—	*Bacillus* Cp-6	1·9	6·6
B. pumilus	5·1	9·7	*Bacillus* Cp-7	6·3	8·4
B. megaterium	5·2	—	*Bacillus* Cp-12	5·6	8·0
B. coagulans	5·5	—	*Bacillus* G-10A	1·2	8·3
B. cereus	5·8	9·4	*Bacillus* GW-21	3·5	7·2
B. licheniformis	6·2	—	*B. psychrosaccharolyticus*	4·5	8·5
B. laterosporus	6·4	10·1	*B. globisporus*	11	7·8
Bacillus DPL	6·6	11·0	*B. macquariensis*	2·5	—
Cl. hastiforme	1·6	11·5			

[a] From Bhadsavle *et al.* (1972); Shehata and Collins (1972); Michels and Viser (1976).

sporeformers, vegetative bacteria (such as Salmonellae) and yeasts and mould spores. The steel screw-cap tubes and the stirring device are most advantageously used in studies with high temperature exposure and short heating times.

Acknowledgements

The author wishes to acknowledge the major contribution of Mr. R. F. Kagei to the development of the stirring device for screw-cap tubes described in this paper.

References

ALDERTON, G., CHEN, J. K. & ITO, K. A. (1974). Effect of lysozyme on the recovery of heated *Clostridium botulinum* spores. *Applied Microbiology* **27,** 613–615.

BHADSAVLE, C. H., SHEHATA, T. E. & COLLINS, E. B. (1972). Isolation and identification of psychrophilic species of *Clostridium* from milk. *Applied Microbiology* **24,** 699–702.

DELPHIN, K. A. (1971). Nacka: ein rationelles Verpflegungssystem. *Hotel- und Gastgewerbe-Rundschau* **11,** 704–705.

GROSSKOPF, J. C. & HARPER, W. J. (1969). Role of psychrophilic sporeformers in long life milk. *Journal of Dairy Science* **52,** 897.

GROSSKOPF, J. C. & HARPER, W. J. (1974). Isolation and identification of psychrotrophic sporeformers in milk. *Milchwissenschaft* **29,** 467–470.

HAUSCHILD, A. H. W. & HILSHEIMER, R. (1977). Enumeration of *Clostridium botulinum* spores in meats by a pour-plate procedure. *Canadian Journal of Microbiology* **23,** 829–832.

KOOIMAN, W. J. (1974). The screw-cap tube technique: a new and accurate technique for the determination of the wet heat-resistance of bacterial spores. In

Spore research 1973 (Barker, A. N., Gould, G. W. and Wolf, J., eds). London and New York: Academic Press, pp. 87–92.

KOOIMAN, W. J. & GEERS, J. M. (1975). Simple and accurate technique for the determination of heat resistance of bacterial spores. *Journal of Applied Bacteriology* **38,** 185–189.

LAINE, J. J. (1970). Studies on psychrophilic Bacilli of food origin. *Annales Academiae Scientiarum Fennicae, Series A, IV Biologica* **169,** 1–36.

LANGEVELD, L. P. M., CUPERUS, F. & STADHOUDERS, J. (1973). Bacteriological aspects of the keeping quality at 5°C of reinfected and non-reinfected pasteurized milk. *Netherlands Milk and Dairy Journal* **27,** 54–65.

LARKIN, J. M. & STOKES, J. L. (1966). Isolation of psychrophilic species of Bacilli. *Journal of Bacteriology* **91,** 1667–1671.

MARSHALL, B. J. & OHYE, D. F. (1966). *Bacillus macquariensis n.sp.*, a psychrotrophic bacterium from sub-antarctic soil. *Journal of General Microbiology* **44,** 41–46.

MICHELS, M. J. M. & VISSER, F. M. W. (1976). Occurrence and thermoresistance of spores of psychrophilic and psychrotrophic aerobic sporeformers in soil and foods. *Journal of Applied Bacteriology* **41,** 1–11.

MOURGUES, R. & AUCLAIR, J. (1973). Durée de conservation à 4°C et 8°C du lait pasteurisé conditionné aseptiquement. *Le Lait* **53,** 481–490.

SHEHATA, T. E. & COLLINS, E. B. (1971). Isolation and identification of psychrophilic species of *Bacillus* from milk. *Applied Microbiology* **21,** 466–469.

SHEHATA, T. E. & COLLINS, E. B. (1972). Sporulation and heat resistance of psychrophilic strains of *Bacillus*. *Journal of Dairy Science* **55,** 1405–1409.

SHEHATA, T. E., DURAN, A. & COLLINS, E. B. (1971). Influence of temperature on the growth of psychrophilic strains of *Bacillus*. *Journal of Dairy Science* **54,** 1579–1582.

SINCLAIR, N. A. & STOKES, J. L. (1964). Isolation of obligately anaerobic psychrophilic bacteria. *Journal of Bacteriology* **87,** 562–565.

STUMBO, C. R. (1965). *Thermobacteriology in food processing*. London and New York: Academic Press.

TÄNDLER, K. (1972). In Folienbehaltnissen pasteurisierte Fleischfertiggerichte für die Gemeinschaftsverpflegung. Teil II: Herstellung und Lagerstabilität der Fertiggerichte. *Die Fleischwirtschaft* **52,** 1105–1113.

WASHAM, C. J., OLSON, H. C. & VEDAMUTHU, E. R. (1977). Heat-resistant psychrotrophic bacteria isolated from pasteurized milk. *Journal of Food Protection* **40,** 101–108.

WINANS, L., PFLUG, I. J. & FOSTER, T. L. (1977). Dry-heat resistance of selected psychrophiles. *Applied and Environmental Microbiology* **34,** 150–154.

An Inflatable Anaerobic Glove Bag

J. W. LEFTLEY AND I. VANCE*

Dunstaffnage Marine Research Laboratory, Oban, Argyll, Scotland

Introduction

During studies on the microbiology of anoxic sediments in Loch Eil, a Scottish sea-loch, it became necessary to develp an isolator in which sediment cores 245 × 64 mm could be manipulated under an inert atmosphere and at low temperatures. The temperature of the bottom water and sediment in Loch Eil varies between 6°C (winter) and 12°C (summer). The isolator had to be easily portable so that it could be assembled in a constant-temperature room when required, and reasonably compact when in storage. For these reasons it was decided to adopt an inflatable glove bag design.

A commercially available glove bag (Instruments for Research and Industry, 108 Franklin Avenue, Cheltenham, Pennsylvania 19012, USA) proved unsuitable because it was constructed from thin polyethylene and so had a limited life. The integral gloves were easily punctured and clumsy in use. A PVC gnotobiotic isolator canopy (type 68023: Vickers Ltd, Medical Engineering, Priestley Road, Basingstoke, Hampshire RG24 9NP) proved to be a more durable alternative when modified for the present application.

The final design (Fig. 1), as described here, allows two operators to work simultaneously, which is necessary when sectioning sediment cores. The complete system is inexpensive to construct (less than £80 at 1977 prices); the canopy may be replaced, if necessary, for less than £20. A similar system, described by Aranki and Freter (1972), is marketed by Coy Laboratory Products Inc. (PO Box 1108, Ann Arbor, Michigan 48106, USA) for $2160 (1977 prices).

*Present address: Department of Life Sciences, Polytechnic of Central London, London, WIM 8JS, England.

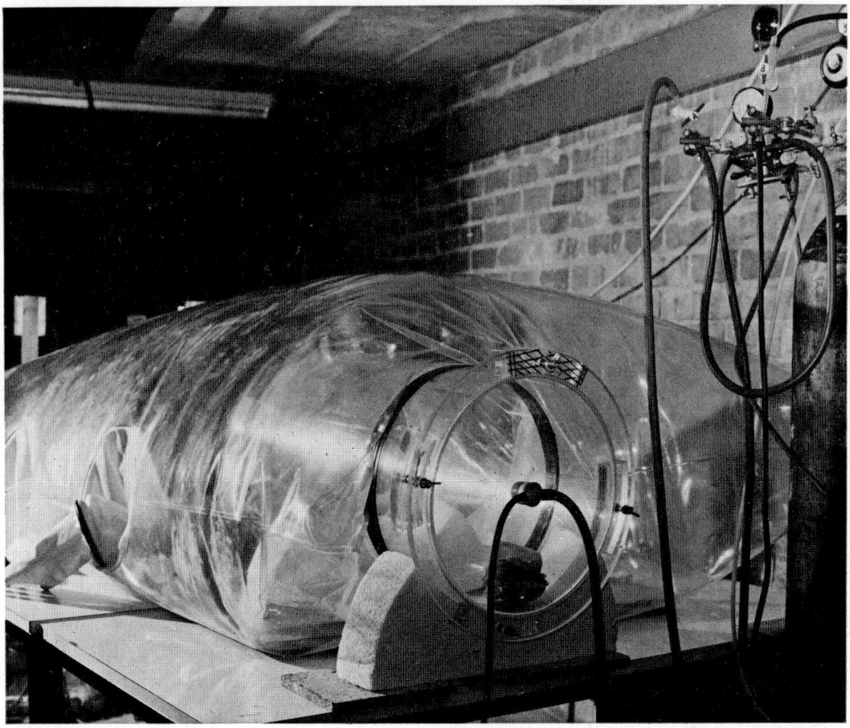

FIG. 1. A general view of the glove bag showing the air lock, gas supply and vacuum system.

Specifications

The canopy

This is made of glass-clear PVC, type LF10, 0·3 mm thick, with sleeves made of 0·3 mm natural frosted PVC. The approximate overall dimensions of the standard canopy when deflated are 2300 × 1200 mm. One end of the bag is unsealed to allow access for cleaning and maintenance. When in use it is sealed by taping the edges together with PVC tape, wrapping the end of the bag several times round a 12 mm diam. "perspex" rod and clamping with "bulldog" clips.

A "formica" covered baseboard, 12 × 750 × 1500 mm, with rounded corners, is placed in the bag to provide an easily cleaned work surface. The board also serves to stabilize the canopy when it is inflated. A few layers of paper towelling are placed between the baseboard and the bottom of the bag, and between the bench top and the lower side of the bag, in order to minimize the risk of puncturing the canopy.

The canopy and airlock can be accommodated on any convenient working surface approximately 1800 × 1200 mm.

In the standard canopy, two pairs of sleeves are located in opposition, midway on the long side, and are on the upper surface when the bag is inflated. The sleeves can be fitted in any desired position to special order and additional sleeves can also be fitted if required. If work is to be done from a seated position it is more convenient to have the sleeves mounted lower than in the standard isolator.

A rubber glove is attached to the end of each sleeve by means of a stainless steel collar and O ring so that punctured gloves can be replaced easily and without deflating the canopy.

The airlock

This can be of any required design, material and dimensions, to suit the needs of the user. It should be able to withstand an internal absolute pressure from 0 to 136 kPa (1 atmosphere = 101 kPa). The airlock can be

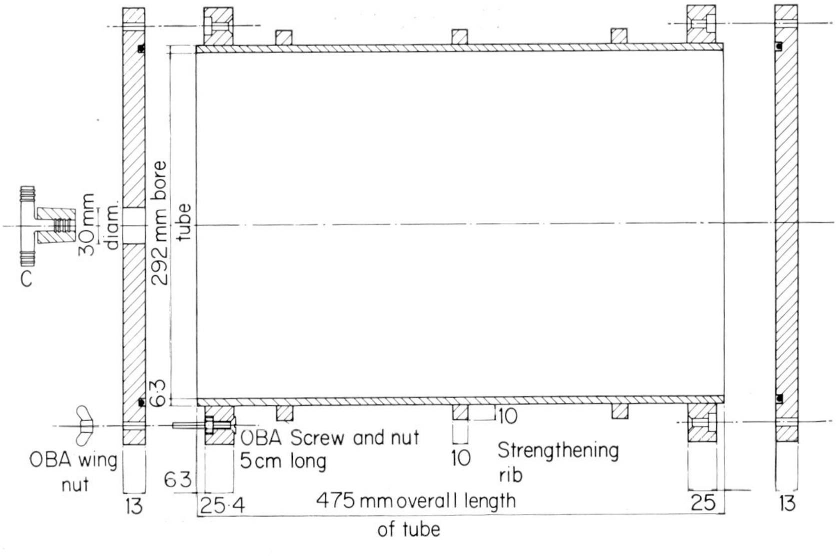

FIG. 2. Plan of the airlock. This is made from 6 mm wall spun acrylic tubing. The doors are machined from 13 mm "perspex" and fitted with O ring seals. C, connector to gas/vacuum supply. Dimensions are in mm unless otherwise indicated.

located in any convenient position but should be placed so that the interior is easily accessible to at least one pair of gloves.

The authors have used a clear acrylic plastic cylinder (Stanley Plastics Ltd, Midhurst, Sussex GU29 9HX), as shown in Fig. 2. It has been found necessary to glue strengthening ribs to the cylinder, midway and at both ends, to prevent stress cracks due to flexing of the walls.

The airlock is fitted into the canopy as described by Aranki and Freter (1972). Electrical cables and tubing for inert gas supply or aspiration can easily be passed through the walls of the canopy using this method of sealing.

Inert gas supply

Cylinders of oxygen-free nitrogen (OFN) normally contain up to 10 $\mu l\ l^{-1}$ of residual oxygen. The OFN supply to the canopy is further purified by passage over BASF Catalyst R3-11 (Anon. 1971: BASF UK Ltd, PO Box 4, Earl Road, Cheadle Hulme, Cheshire SK8 6QG) maintained at 160°C. The gas supply system normally operates at a working pressure of 21 kPa and a flow rate of 10 l min^{-1}. Butyl rubber, which is impermeable to oxygen, is used for the gas lines.

Vacuum supply

A good water pump will produce a vacuum of about 101 kPa. Alternatively, a diaphragm pump, such as the "Compton" vacuum pump type D/351 VM (Dawson, McDonald and Dawson Ltd, Compton Works, Ashbourne, Derbyshire DE6 1DB), can be used.

Operation

Purging the canopy

The inner door of the airlock is left off and the vacuum supply turned on (B, Fig. 3). The inner surfaces of the canopy are drawn closely together and cling tightly round the airlock. The bolts on the airlock are covered with pads of rubber to prevent the canopy being punctured as it draws tightly round the mouth of the airlock. Valve B is then closed and valve A opened to admit nitrogen. In the other position, valve A vents into the air to prevent back-pressure in the gas scrubber. During normal operation the valve feeds into the by-pass line which is connected directly into the canopy. When the bag is partly reflated the gas supply is shut off and the bag re-evacuated. This cycle of evacuation and inflation is repeated

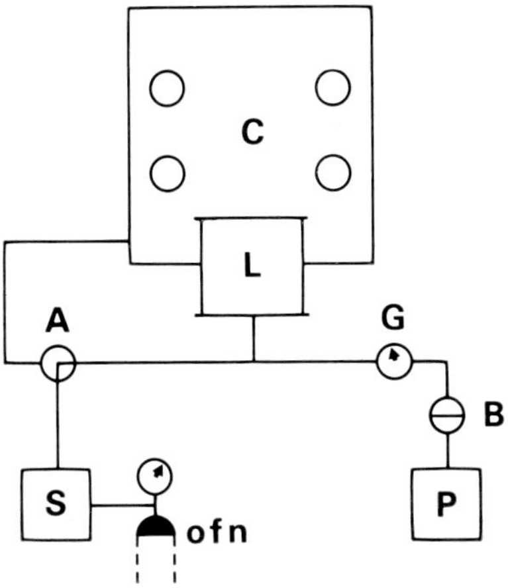

FIG. 3. Schematic diagram of the glove bag system. A, 2-way valve for gas supply; B, valve for vacuum supply; C, canopy; G, vacuum gauge; L, air lock; P, vacuum pump; S, gas purifier; ofn, cylinder of oxygen-free nitrogen.

several times and then the canopy is inflated until it stands about 550 to 600 mm high. The inner airlock door is secured and the bag is then ready for use. Purging of the bag normally takes about 20 min.

Normal operation

The airlock is loaded with equipment and then purged in the same way as described above. This operation takes about 5 min. Items of equipment may thus be moved in and out of the glove bag relatively quickly. When small items are being transferred a "dummy load", as described by Aranki and Freter (1972), may be used to reduce the gas volume in the lock.

If it is to be used frequently, the canopy can be left inflated and it will remain so for at least a week without further pressurization, provided that there are no leaks in the system. When the bag is not in use a container of alkaline pyrogallol can be left inside to absorb any residual oxygen.

Applications

We do not claim that the system described can provide an absolutely anaerobic environment. Trace amounts of oxygen have not been a problem in our work with sulphide-rich sediments which have sufficient buffering capacity, with respect to sulphide, to protect anaerobes from the effects of traces of oxygen. Intact sediment cores are quickly cut up,

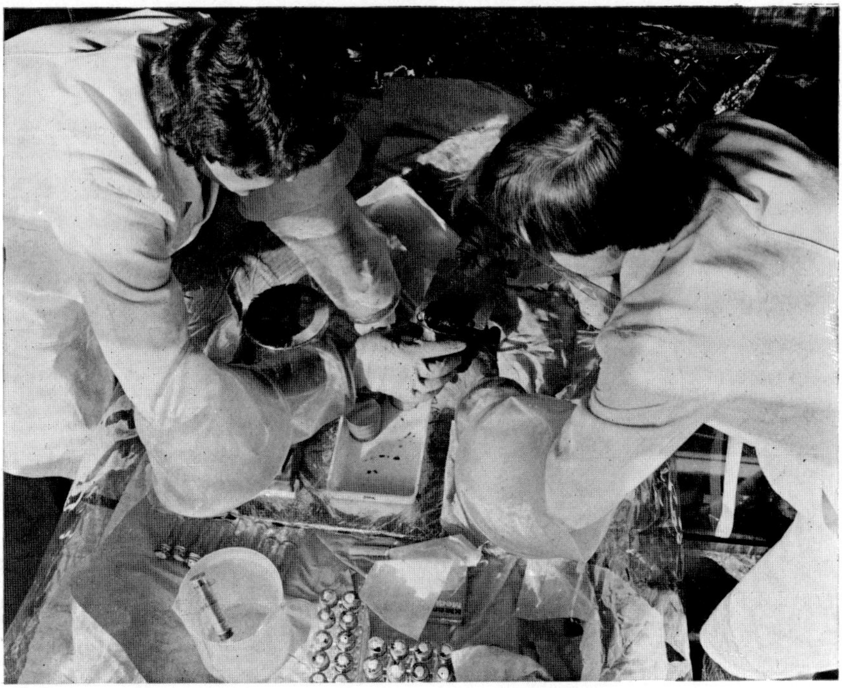

FIG. 4. Sediment cores being cut up in the bag.

dispensed into the appropriate anaerobic culture medium and placed in anaerobic jars (Fig. 4).

It is suggested that for work with anaerobes such as methanogenic bacteria an ultra-low oxygen chamber, as described by Edwards and McBride (1975), could be used inside the bag.

The canopy has also proved useful for work on the chemistry of the interstitial water of anoxic sediments where oxidation during handling must be avoided (Troup *et al.*, 1974).

It would be possible to make this system more sophisticated by incorporating such refinements as continuous recirculation of the gas within the canopy through the gas scrubber, improved airlock design and an automatic valve to maintain the canopy at a constant pressure.

Acknowledgements

I.V. acknowledges the receipt of a studentship from the Natural Environment Research Council. We thank Mr. K. Hoare for drawing Fig. 2.

References

ANON. (1971). *BASF Catalyst R3-11, Technical Leaflet.*

ARANKI, A. & FRETER, R. (1972). Use of anaerobic glove boxes for the cultivation of strictly anaerobic bacteria. *American Journal of Clinical Nutrition* **25,** 1329–1334.

EDWARDS, T. & MCBRIDE, B. C. (1975). New method for the isolation and identification of methanogenic bacteria. *Applied Microbiology* **29,** 540–545.

TROUP, B. N., BRICKER, O. P. & BRAY, J. T. (1974). Oxidation effect on the analysis of iron in the interstitial water of recent anoxic sediments. *Nature, London* **249,** 237–239.

Alteromonas (Pseudomonas) putrefaciens

J. V. LEE

Public Health Laboratory, Preston Hall Hospital, Maidstone, Kent, England

Introduction

Derby and Hammer (1931) first described *Alteromonas* (*Pseudomonas*) *putrefaciens* as the organism responsible for surface taint, a cheesy or putrid condition of salted butter. It was first classified as an *Achromobacter* sp. (Derby and Hammer, 1931) but later transferred to the genus *Pseudomonas* (Long and Hammer, 1941a). For reasons discussed later this classification is unsatisfactory, and this led Lee *et al.* (1977) to transfer it to the genus *Alteromonas*.

Sources and Significance

Alteromonas putrefaciens is extremely widely distributed as is shown by the range of environments from which it has been isolated. These include raw fresh milk, raw fresh cream, normal and putrid salted butter, moist soil, streams, lakes, roadside water, sewers (Long and Hammer, 1941b), fish (Castell *et al.*, 1949; Chai *et al.*, 1968), seawater (Lee *et al.*, 1977), estuarine muds, shellfish (Lee, unpublished results), poultry (Barnes and Impey, 1968), natural gas and oil brines (Iizuka and Komagata, 1964), cutting oil (Pivnick, 1955) and various clinical sites including the ear, sputum, leg ulcers, urine, faeces, and tonsillar pus (Minagawa, 1963; von Graevenitz and Simon, 1970; Gilardi, 1972; Levin, 1972; Riley *et al.*, 1972; Debois *et al.*, 1975; Holmes *et al.*, 1975).

Alteromonas putrefaciens is probably of most importance in the food industry. Its ability to grow at chill room temperatures (0–7°C), to break down a wide range of compounds external to the cell (see below), to produce malodorous compounds such as hydrogen sulphide and trimethylamine, and its wide distribution make it potentially of very great importance in the spoilage of proteinaceous foods at low temperatures.

Thus, *A. putrefaciens* was shown to constitute up to 10% of the total bacterial flora of chicken carcass after 10–11 d at 1°C (Barnes and Thornley, 1966). If the chicken carcasses are wrapped in oxygen-impermeable film the proportion may be as high as 69% of the total aerobic bacterial flora after 12 d at 1°C (Barnes and Melton, 1971). It is also one of the organisms most commonly isolated from fish spoiling at 0–4°C (Castell *et al.*, 1949; Chai *et al.*, 1968). Shewan (1974) suggested that *A. putrefaciens* may constitute at least one third of the total flora of foods such as poultry, fish and possibly meat when spoilage occurs. He is of the opinion that this is one of the most important organisms encountered in the spoilage of protein foods kept at chill-room temperatures.

Although Pivnick (1955) suggested that it might have a role in the biodegradation of cutting oils, the significance of the presence of *A. putrefaciens* in oil/water emulsions has not been determined. Similarly, although it has been isolated from the many types of clinical specimen listed above its pathogenicity has not been proven. Holmes *et al.* (1975) suggest that it may have a pathogenic role in ear infections and it certainly appears to have been the main infective agent in the case of the leg ulcer described by Debois *et al.* (1975). None the less, it is likely to be encountered only rarely as an opportunistic pathogen particularly and then only in patients whose immune response is debilitated.

Description and Characteristics of *Alteromonas putrefaciens*

Alteromanas putrefaciens is a Gram negative rod-shaped bacterium 0·5–1·0 by 1·1–4·0 μm, which occurs singly or in pairs and is motile by a single polar flagellum. Colonies of the organism on nutrient agar possess a pinkish or reddish-brown pigment, are 2–3 mm diam., domed and glistening. Colony variation may occur if repeatedly subcultured. The optimal temperature for growth is usually in the range 20–25°C, although strains isolated from clinical material may grow at 37°C or even 42°C. Strains isolated from chilled foods are frequently capable of growth at 0°C but not at 37°C. Using a temperature gradient apparatus it has been shown that strains may grow at as low as —4°C and in one case down to —6°C (D. M. Gibson, pers. comm.).

Some strains require Na^+ for growth although at very low concentrations (Riley *et al.*, 1972; Lee, 1973) and the growth of many strains is inhibited in the absence of Na^+. No growth factors are required and all strains are capable of growth on a simple mineral salts medium containing succinate as the sole source of organic carbon. The range of compounds supporting growth is limited but most strains can grow on glucose, glucosamine, *N*-acetylglucosamine, isoleucine, threonine, acetate, propionate, lactate, pyruvate, fumarate and malate (Lee *et al.*, 1977).

In Hugh and Leifson's (1953) test for the metabolism of glucose the majority of strains give either no reaction or alkali production in the open tube. Some strains may give a weak oxidative reaction but they are never fermentative. All strains examined have been shown to possess the key enzymes of the Entner-Doudoroff pathway, 6-phosphogluconate dehydrase, and 2-keto-3-deoxy-6-phosphogluconate aldolase. They also possess dehydrogenase activity on 6-phosphogluconate and glucose-6-phosphate which is perhaps indicative of the pentose phosphate pathway (Lee, 1973; Lee *et al.*, 1977). As noted above most strains can grow on glucose. Thus although they normally show no signs of the metabolism of glucose in Hugh and Leifson's (1953) test they can metabolize glucose probably by the Entner-Doudoroff and/or the pentose phosphate pathways.

All strains are catalase and Kovacs (1956) oxidase positive, reduce nitrate to nitrite and trimethylamine oxide to trimethylamine (Lee *et al.*, 1977). Most strains are strong producers of H_2S (Levin 1968), although some strains from clinical specimens have been reported as negative or very weak producers (Levin, 1975). These weak or negative H_2S producers can give variants that are strong H_2S producers (Levin, 1975). Nearly all strains produce ornithine but not lysine decarboxylase or arginine dihydrolase.

In addition to the ability to produce H_2S and trimethylamine one of the most striking characteristics of *A. putrefaciens* is the wide range of compounds that it can attack extracellularly. This range includes gelatin, casein, "Azocoll" (a degraded form of collagen), albumin, deoxyribonucleic acid (DNA), ribonucleic acid, lipids, lecithin, chitin and phosphates (Sadovski and Levin, 1969; Hugh, 1970; Barnes and Melton, 1971).

Isolation and Identification

Enrichment broths and selective media are neither necessary nor available for the isolation of this organisms. It will not grow on the cetrimide-containing media sometimes used for the isolation of *Ps. aeruginosa*. If *A. putrefaciens* is present in significant numbers in a sample it will be detected by direct plating on nutrient agar. Obviously if the bacterial count is high, serial dilutions of a suspension of the material being examined may be necessary. Nutrient agar plates are incubated for 24–48 h at 20–25°C when *A. putrefaciens* will appear as smooth glistening pinkish to reddish-brown colonies of 1–3 mm.

The minimum characters for the identification of *A. putrefaciens* outlined by Hugh (1970) are suitable for the preliminary identification. These characters together with some others that might be useful for identification are shown in Table 1.

TABLE 1. Characters useful for the identification of *A. putrefaciens*

Property	Results
Gram stain	Gram negative rods
Motility	+
Flagella	Polar monotrichous
Pigment (nutrient agar)	Salmon pink/reddish-brown
Hugh and Leifson's test	
open tube	NC[a]/alkaline (occasionally acid)
sealed tube	—
Kovacs oxidase	+
Catalase	—
H_2S (Kligler iron agar, etc.)	+
Lysine decarboxylase	—
Ornithine decarboxylase	+
Arginine dihydrolase	—
Gelatinase	+

[a] NC, no change.

Taxonomic Considerations

This group of bacteria after initially being identified as an *Achromobacter* sp. (Derby and Hammer, 1931), was placed in the genus *Pseudomonas* (Long and Hammer, 1941a). Its identification as *Pseudomonas* sp. was not questioned until the introduction of DNA-base composition determination as a taxonomic tool. *Pseudomonas* is now recognized as having a mol % guanine-cytosine (GC) of 58–72 (Doudoroff and Palleroni, 1974). The mol % GC of *A. putrefaciens* strains is too low for *Pseudomonas*, being in the range 43–54 (Lee *et al.*, 1977; Owen *et al.*, 1978). This range ignores the figures of Levin (1972) because they are consistently higher than those of other workers by 3–4 mol % GC as pointed out by Lee (1973) and Owen *et al.* (1978). This is because Levin used a value of 54·0 mol % GC for the DNA of the reference strain *Eschericia coli* strain B whereas other workers have taken the value to be 50·9 (Owen *et al.*, 1978).

Lee *et al.* (1977) carried out a numerical taxonomic study on pseudomonas-like bacteria using biochemical data from rapid screening assays for dehydrogenases in cell fractions, and other more usual taxonomic data such as nutritional screening. They concluded that *Ps. rubescens* is a synonym of *Ps. putrefaciens*, as had been suggested by Hugh (1970), and that the species should be transferred from *Pseudomonas* to the genus *Alteromonas* recently described by Baumann *et al.* (1972). *Alteromonas* was created to include marine pseudomonad-like bacteria that differed from *Pseudomonas* in having a mol % GC of 43–48.

The range 43–54 mol % GC of *A. putrefaciens* is larger than that normally accepted for a single species. Levin (1972) claimed that there were two groups within the putrefaciens complex. The members of group 1 have low (47·8–50·8) mol % GC and could not grow in broth containing 6% NaCl whereas those of group 2 have high (55·9–59·0) mol % GC and can grow in 6% NaCl. Riley *et al.* (1972) described two similar groups but did not report their DNA base compositions. Their group 1 strains in addition to being unable to grow in 6% NaCl were also unable to grow on *Salmonella-Shigella* (SS) agar and produced acid from sucrose, maltose, arabinose and glucose in OF (oxidative-fermentative) medium. Their group 2 were unable to grow in the absence of NaCl but grew at concentrations of 7–10%, were able to grow on SS agar and unable to produce detectable amounts of acid from any carbohydrates tested. These groupings were largely confirmed by Holmes *et al.* (1975). Although it might be expected that such differences would reflect the origins of the strains this does not appear to be the case. Contrary to expectations, many fisheries' isolates are apparently able to grow in the absence of NaCl (Riley *et al.*, 1972; Holmes *et al.*, 1975).

Most recently, Owen *et al.* (1978) examined the DNA homology of strains of *A. putrefaciens* and their results suggest that there are four groups. The low mol % GC strains fall into three DNA homology groups. Group I is of relatively high internal DNA relatedness, whereas groups II and III are of relatively low internal relatedness. The remaining group IV includes all the high mol % GC strains and has high levels of internal relatedness. The one character distinguishing the high and low mol % GC strains is growth in the presence of 6% NaCl, but various other phenotypic traits were suggested as being useful for distinguishing the four DNA-homology groups (Owen *et al.*, 1978).

From the above discussion it may be concluded that there are at least two distinct subgroups within the putrefaciens complex. Although the groups I and IV of Owen *et al.* (1978) by themselves appear distinct enough to warrant the status of separate species the positions of the intermediate groups II and III are doubtful. It would therefore seem wisest not to divide the putrefaciens group into separate species or biotypes at present. It is certain that it does not belong to the genus *Pseudomonas*, but its position in the genus *Alteromonas* should be confirmed or refuted by DNA-homology studies. Notwithstanding the problems concerning the taxonomy of *A. putrefaciens*, which are likely to occupy taxonomists for some years, it remains a very easy group of organisms to identify. Finally, there can be little doubt of its importance in the spoilage of proteinaceous foods at temperatures of 0–7°C.

References

BARNES, E. M., & IMPEY, C. S. (1968). Psychrophilic spoilage bacteria of poultry. *Journal of Applied Bacteriology* **31,** 97–107.

BARNES, E. M. & MELTON, W. (1971). Extracellular enzymic activity of poultry spoilage bacteria. *Journal of Applied Bacteriology* **34,** 599–609.

BARNES, E. M. & THORNLEY, M. J. (1966). The spoilage flora of eviscerated chickens stored at different temperatures. *Journal of Food Technology* **1,** 113–119.

BAUMANN, L., BAUMANN, P., MANDELL, M. & ALLEN, R. D. (1972). Taxonomy of aerobic marine eubacteria. *Journal of Bacteriology* **110,** 402–429.

CASTELL, C. H., RICHARDS, J. F. & WILMOT, I. (1949). *Pseudomonas putrefaciens* from cod fillets. *Journal of the Fisheries Research Board of Canada* **7,** 430–431.

CHAI, T., CHEN, C., ROSEN, A., & LEVIN, R. E. (1968). Detection and incidence of specific species of spoilage bacteria on fish. II. Relative incidence of *Pseudomonas putrefaciens* and fluorescent pseudomonads on haddock fillets. *Applied Microbiology* **16,** 1738–1741.

DEBOIS, J., DEGREEF, H., VANDEPITTE, J. & SPAEPEN, J. (1975). *Pseudomonas putrefaciens* as a cause of infection in humans. *Journal of Clinical Pathology* **28,** 993–996.

DERBY, H. A. & HAMMER, B. W. (1931). Bacteriology of butter. IV. Bacteriological studies on surface taint butter. *Iowa Agricultural Experimental Station Research Bulletin* **145,** 389–416.

DOUDOROFF, M. & PALLERONI, N. J. (1974). Genus I. *Pseudomonas* Migula 1894. In *Bergey's manual of determinative bacteriology* (Buchanan, R. E. & Gibbons, N. E., eds) 8th edn. Baltimore: Williams and Wilkins, p. 217.

GILARDI, G. L. (1972). Infrequently encountered *Pseudomonas* species causing infections in man. *Annals of Internal Medicine* **77,** 211–215.

HOLMES, B., LAPAGE, S. P. & MALNICK, H. (1975). Strains of *Pseudomonas putrefaciens* from clinical material. *Journal of Clinical Pathology* **28,** 149–155.

HUGH, R. (1970). A practical approach to the identification of certain nonfermentative Gram negative rods encountered in clinical specimens *Public Health Laboratory* **28,** 168–187.

HUGH, R. & LEIFSON, E. (1953). The taxonomic significance of fermentative versus oxidative metabolism of carbohydrates by various Gram negative bacteria. *Journal of Bacteriology* **66,** 24–26.

IIZUKA, H. & KOMAGATA, K. (1964). Microbiological studies on petroleum and natural gas. II. Determination of pseudomonads isolated from oil brines and related materials. *Journal of General and Applied Microbiology, Tokyo* **10,** 223–231.

KOVACS, N. (1956). Identification of *Pseudomonas pyocyanea* by the oxidase reaction. *Nature (London)* **178,** 703.

LEE, J. V. (1973) *Some comparative biochemical and physiological studies on selected Gram-negative bacteria.* Ph.D. thesis, University of Aberdeen.

LEE, J. V., GIBSON, D. M. & SHEWAN, J. M. (1977). A numerical taxonomic study of some Pseudomonas-like marine bacteria. *Journal of General Microbiology* **98,** 439–451.

LEVIN, R. E. (1968). Detection and incidence of specific spoilage bacteria on fish. I. Methodology. *Applied Microbiology* **16,** 1734–1737.

LEVIN, R. E. (1972). Correlation of DNA base composition and metabolism of *Pseudomonas putrefaciens* isolates from food, human clinical specimens, and other sources. *Antonie van Leeuwenhoek* **38,** 121–127.

LEVIN, R. E. (1975). Characteristics of weak H_2S-producing isolates of *Pseudomonas putrefaciens* from human infections. *Antonie van Leeuwenhoek* **41,** 569–574.

LONG, H. F. & HAMMER, B. W. (1941a). Classification of organisms important in dairy products. III. *Pseudomonas putrefaciens. Iowa Agricultural Experimental Station Research Bulletin* **285,** 176–195.

LONG, H. F. & HAMMER, B. W. (1941b). Distribution of *Pseudomonas putrefaciens. Journal of Bacteriology* **41,** 100–101.

MINAGAWA, M. (1963). Studies on the strains closely related to *Vibrio parahaemolyticus* and reddish-brown pigment-producing Pseudomonas isolated from the stools of patients with acute enteritis. *Annual Report of the Institute of Food Microbiology* (*Chiba University*) **16,** 9–23.

OWEN, R. J., LEGROS, R. M. & LAPAGE, S. P. (1978). Base composition, size and sequence similarities of genome deoxyribonucleic acids from clinical isolates of *Pseudomonas putrefaciens. Journal of General Microbiology* **104,** 127–138.

PIVNICK, H. (1955). *Pseudomonas rubescens*, a new species from soluble oil emulsions. *Journal of Bacteriology* **70,** 1–6.

RILEY, P. S., TATUM, H. W. & WEAVER, R. E. (1972). *Pseudomonas putrefaciens* isolates from clinical specimens. *Applied Microbiology* **24,** 798–800.

SADOVSKI, A. Y. & LEVIN, R. E. (1969). Extracellular nuclease activity of fish spoilage bacteria, fish pathogens, and related species. *Applied Microbiology* **17,** 787–789.

SHEWAN, J. M. (1974). The biodeterioration of certain proteinaceous foodstuffs at chill temperatures. In *Industrial aspects of biochemistry* (Spencer, B., ed.). London: North-Holland Publishing Company.

VON GRAEVENITZ, A. & SIMON, G. (1970). Potentially pathogenic, non-fermenta tive, H_2S-producing Gram negative rod (Ib). *Applied Microbiology* **19,** 176.

Techniques Used for Studying Terrestrial Microbial Ecology in the Maritime Antarctic

D. D. WYNN-WILLIAMS

Life Sciences Division, British Antarctic Survey, Madingley Road, Cambridge, Cambridgeshire, England

Introduction

Many years have passed since the pioneering of antarctic microbiology by Ekelöf (1908), Tsiklinsky (1908) and Pirie (1912), but although sophisticated techniques are now used at antarctic research stations the basic problems of aseptic sampling and treatment of peat and soils at low temperatures remain. Coring frozen peat in winter demands different procedures from the sampling of delicate moss associations in summer, and studies of the crucial freeze-thaw transformation call for modifications to standard procedures.

The objectives of the present research at the Signy Island terrestrial reference sites (SIRS) and related areas were to determine the size and fluctuations of bacterial, yeast and mould populations in peat relative to decomposer activity. This was assessed by measurement of oxygen uptake of whole cores and by assay of loss in tensile strength of cotton strips. The work was part of a long-term study of the flora and fauna responsible for production and decomposition at the SIRS (Tilbrook, 1973) and followed two earlier microbiological studies on the island: Bailey and Wynn-Williams (in press), Baker (1970a, b) and Baker and Smith (1972). The methodology was developed with reference to the bipolar studies of the International Biological Programme, IBP (Rosswall, 1971; Rosswall and Heal, 1975).

Selection of Sampling Areas

Several locations are being studied by British Antarctic Survey microbiologists in the maritime antarctic. They are Rothera Point (67° 34′ S,

68° 08′ W) in Adelaide Island; SIRS (60° 43′ S, 45° 36′ W) and Lynch Island (60° 39′ S, 45° 36′ W), both in the South Orkney Islands; Maiviken (54° 14′ S, 36° 30′ W) in South Georgia; and Bird Island (54° 00′ S, 38° 03′ W).

The procedures reported here were developed mainly for decomposition studies at SIRS which are described by Tilbrook (1973) and Collins *et al.* (1975). These consist of a relatively dry moss turf community (SIRS 1) dominated by *Polytrichum alpestre* and *Chorisodontium aciphyllum*, and a wet moss carpet community (SIRS 2) dominated by *Calliergon sarmentosum* and *Cephaloziella varians* with patches of *Drepanocladus uncinatus* and *Calliergidium austro-stramineum*.

Collection of Samples

Spring and summer

For comparative physiological purposes, peat cores from pure moss stands are required. This entails stratified random sampling within each reference site and adequate replication to ensure statistical significance of both microbial and respirometric data. The fragile nature of the mosses and the underlying peat results in compression and drainage during coring with subsequent desiccation and changes in both temperature and aeration, which must be minimized. Robust, sterilizable containers have to be used for fragile summer cores, but rigid frozen winter cores need less protection.

Cores were located using a 1 m^2 wire quadrat subdivided into 10 cm squares, and were collected 20 cm apart. Random numbers were employed but squares not containing the moss in sufficient cover were rejected. Four replicate cores m^{-2} quadrat were taken for microbial assessment and eight for Gilson respirometry. To minimize distortion of the moss and peat, an ethanol-rinsed, flame-sterilized steel corer with a 12 × 1 cm sharpened groove (after Michelbacher, 1939) was employed. It was transported with the grooved cutting end encased in the sterile outer container of a 60 ml Brunswick catheter syringe (Southern Syringes Services Ltd, 303 Chase Road, Southgate, London NW14), and the upper end closed by a sterile aluminium film canister which was left in place during coring. The cores were transferred into autoclaved, sawn-off 60 ml Brunswick catheter syringes, isodiametric (27 mm internal diam., cross-sectional area 573 mm^2) with both the corer and the Quickfit B34/35 cones used to hold core sections for respirometry. A sterile syringe-piston and ramrod were used to transfer the core, the groove being sealed by inserting the corer in a sawn-off syringe-container

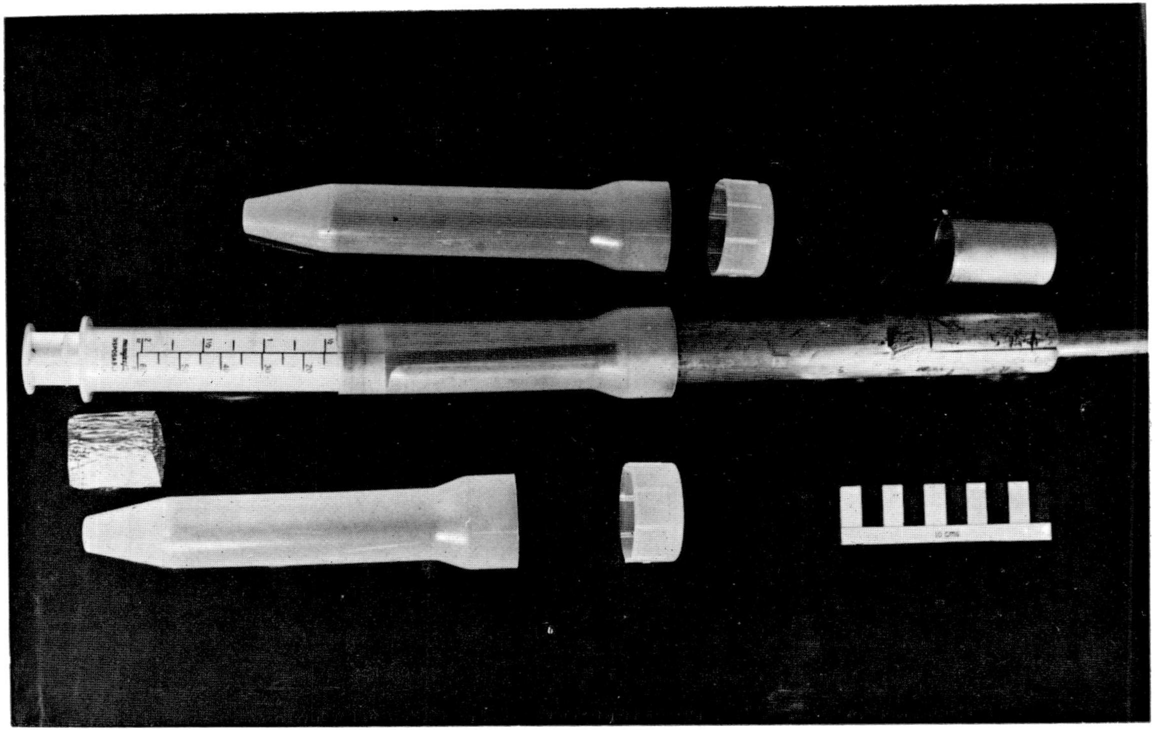

FIG. 1. Equipment used for aseptic coring of Antarctic peat in summer. The scale-marker is 10 cm long. For information, see text.

(Fig. 1). The full syringes, minus their plungers, were inserted in their containers and capped for aseptic transport to the laboratory in a Dewar flask or a "space blanket"-wrapped box containing ice-packs. Transport and treatment were as rapid as possible to minimize microbial changes after sampling, the entire procedure being completed within 12 h.

Winter

The snow depth is up to 1·0 m on SIRS 1 and 1·5 m on SIRS 2, with up to 15 cm depth of ice underneath. A compromise between adequate exposure of the moss surface and prevention of damage has to be made. The frozen peat is extremely hard making drilling necessary. This together with the hazards imposed by winter clothing (Fig. 2) increases the risk of contamination.

The surface was exposed using a clean shovel and an axe and sampled with a carbide-toothed corer on a power drill driven by a field generator. The corer was ethanol-rinsed and flamed in the laboratory before use; flaming in the field was impossible. Cores frequently froze into the corer which was then warmed near the generator exhaust before ejecting the cores into ethylene oxide-sterilized (10% v/v in CO_2) polythene bags sealed with pipe cleaners or into Brunswick syringes. A Dewar flask or ice packs prevented cores from thawing *en route* to the laboratory.

Treatment of Samples

Subsampling

Minimization of temperature changes and contamination are major considerations. Cores were stored at field temperature at the time of collection in incubators or constant-temperature rooms for as short a time as possible. Laboratory air temperature was kept below 10°C, and a large cooled inverted aluminium tray at *ca.* 3°C served as a cutting and working surface. This was ethanol-rinsed and flamed and used for cooling flamed instruments, bottles and plates. A Resiguard (Nicholas Laboratories Ltd, Slough, UK) spray was used to settle dust and a face mask was worn during vigorous sectioning of frozen cores. In summer, a sterile knife with a long, thin curved blade proved ideal for sectioning delicate cores braced between sterile syringes. In winter a hacksaw blade (7 teeth cm^{-1}) was satisfactory. The green upper 1 cm of the core was removed, and sections 1–3, 3–6, 6–9 and 9–12 cm were cut or subsampled as required. Three subsamples per section, two for microbial assessment and one for water content determination, were taken using a

FIG. 2. Drilling peat cores in winter at Signy Island, illustrating contamination problems.

flamed 8 mm cork-borer to obtain 1–2 g fresh weight subsamples. These were weighed before homogenizing in 99 ml volumes of diluent containing 0·1 g peptone and 0·2 g sodium hexametaphosphate 100 ml^{-1} in ethanol-rinsed and flamed ice-cooled MSE vortex beakers for 2 min at the maximum revolutions of an MSE homogenizer.

Spread plate preparation

Nystatin (E. R. Squibb and Sons, Moreton, Merseyside; 50 mg l^{-1}) and Actidione (The Upjohn Company, Michigan, USA; 50 mg l^{-1}) were added to bacterial media (Williams and Gray, 1973), which included casein peptone starch agar (CPSA), nutrient agar (Difco, NAD), nutrient agar (Oxoid, NAOx), tryptone soya agar (TSA), $\frac{1}{10}$ strength TSA (TSA $\frac{1}{10}$), and peat expressate yeast extract agar (PEYEA) as given on p. 73. CPSA was optimal for counts, growth-rate and pigmentation. Sodium propionate (250 mg l^{-1}) and Aureomycin (Chlortetracycline hydrochloride, Lederle Laboratories Division, Cyanamid of Great Britain Ltd, Gosport; 30 mg l^{-1}) were usually added to yeast and filamentous fungal media (Beech and Davenport, 1969; Williams and Gray, 1973), which included Sabouraud dextrose agar (SDA), Czapek Dox yeast extract agar (CDYEA), potato dextrose agar (PDA), glucose peptone agar (GPA) and malt extract agar (MEA) as given on pp. 73–74. SDA was used for routine counting of yeasts and moulds.

Pre-poured plates were air dried for 1–2 d and cooled before use.

Medical flat bottles containing homogenates were shaken 25 times through a distance of at least 0·5 m and settled for 2 min, and dilution series tubes were agitated for 10 s in a vortex mixer. Aliquots of 200 μl were then pipetted aseptically on to triplicate plates by Oxford Sampler (Boehringer Corporation Ltd, Bilton House, Uxbridge Road, Ealing, London W5), and spread using L-shaped glass rods. Routine plates were incubated at 10°C for 14 d before counting, and stored at 0°C for at least 2 weeks before describing the colonies and estimating dominance.

Isolates were picked off the plates into 10 ml volumes of casein peptone starch medium (see p. 73) made with only 0·3% (w/v) agar. After 7 d at 10°C, part of the growth was excised with a sterile loop and suspended in 5 ml volumes of sterile deionized water. The agar cultures were then overlaid with *ca.* 3 ml sterile mineral oil and stored at 2°C for transport to the UK. Thesuspensions were frozen at —10°C to —20°C for transport.

Preparation of slides for direct counts

Aliquots (18 ml) of *ca.* 10^{-2} dilution of homogenate were transferred by sterile syringe into 2 ml volumes of sterile agar diluent (0·1 g 100 ml^{-1}) in McCartney bottles. After shaking, and settling for 2 min, 10 or 20 μl aliquots were transferred by Finnpipette (Jencons Scientific Ltd, Mark Road, Hemel Hempstead, Herts) on to 3·2 mm diam. spots on PTFE-coated Multispot slides (No. SMO10) (C. A. Hendley and Co., Victoria Road, Buckhurst Hill, Essex), and air dried in a laminar flow cabinet or under a dryer. The slides were stored in the dark, cold and dry, prior to staining.

Preparation of cores for Gilson respirometry

Core sections were cut within Brunswick syringes and were inserted aseptically directly into isodiametric ethylene oxide-sterilized Quickfit B34/35 cones lined with paraffin wax to render the sides parallel, thus limiting gas exchange to the upper surface. The cones were either covered with sterile aluminium foil for incubation or screened immediately.

Culture Media and Staining Procedure

All International Biological Programme (IBP) code numbers are from Rosswall (1971).

Media for isolating bacteria from peat

Casein Peptone Starch Agar (NACPSA)

IBP code number: 0200.xxxx.01.08.3

Casein 0·5 g, soluble starch 0·5 g, Bacto peptone (Difco) 0·5 g, glycerol 1·0 ml, yeast extract (Difco) 0·5 g, K_2HPO_4 0·2 g, $MgSO_47H_2O$ 0·05 g, $FeCl_3$ (0·01% aq. w/v) 4 drops, agar 15 g, peat expressate 250 ml, deionized water to 1000 ml.

The soluble starch, casein and phosphate are dissolved separately in 20 ml volumes of deionized water and sterilized at 121°C for 15 min along with the basal medium before aseptic addition at *ca.* 50°C and adjustment of the pH to 6·9.

Nutrient Agar, Difco (NAD)

IBP code number: 0200.xxxx.01.10.3.

Nutrient Agar, Oxoid (NAOx)

Heal *et al.* (1967); Parkinson *et al.* (1971).

Peat Expressate Yeast Extract Agar (PEYEA)

IBP code number: 02000.xxxx.01.08.6.

Peat expressate is diluted with an equal volume of deionized water to which is added 2·0 g l^{-1} yeast extract (Difco) and 15·0 g l^{-1} agar. Adjust the pH to 6·9, and autoclave at 121°C for 15 min.

Tryptone Soya Agar (TSA)

IBP code number: 0200.xxxx.01.04.1.

Media for isolating yeasts and filamentous fungi from peat

Czapek Dox Yeast Extract Agar (CDYEA)

IBP code number: 0100.xxxx.01.03.1.

Enrich with peat expressate (250 ml l^{-1}) and yeast extract (0·5 g l^{-1}) after Latter and Heal (1971).

Glucose Peptone Agar (GPA)

Di Menna (1966). It consists of glucose 40 g, mycological peptone 10 g, agar 20 g, deionized water 1000 ml.

Adjust to pH 4·5 with NHCl. Autoclave at 121°C for 15 min.

Malt Extract Agar (MEA)
Oxoid, CM 59.

Potato Dextrose Agar (PDA)
IBP code number 0105.xxxx.01.10.1.

Sabouraud Dextrose Agar (SDA)
IBP code number 0105.xxxx.01.04.1.

Fluorescein iso-thiocyanate (FITC) staining procedure

Potassium phosphate buffer, pH 7·2

Stock solutions consist of (1) $Na_2HPO_4{\cdot}2H_2O$ 11·876 g l^{-1}, (2) KH_2PO_4 9·078 g l^{-1}.

Mix 7 ml of solution (1) with 3 ml of solution (2) to give pH 7·2. This 0·072 M solution is then diluted to give a final molarity of 0·01.

Sodium carbonate/bicarbonate buffer, pH 9·6.

Mix 2·5 ml of 0·5 M sodium carbonate solution with 7·5 ml of 0·5 M sodium bicarbonate solution.

Final staining solution

This stain is recommended by Babiuk and Paul (1970), and consists of FITC (BBL, Becton Dickinson Ltd) 1 mg, 0·5 M sodium carbonate/bicarbonate buffer (pH 9·6) 0·25 ml, 0·01 M potassium phosphate buffer (pH 7·2) 1·10 ml, 0·85% (w/v) sodium chloride solution 1·00 ml.

Mix thoroughly at room temperature and use immediately. The staining solution should not be stored longer than 6 h at 1°C in the dark.

20 μl of stain is pipetted on to each agar smear on dried Hendley slides and is allowed to dry in a flow of clean air to fix the smears. The slide is then immersed in a trough of sodium carbonate/bicarbonate buffer (pH 9·6) and rocked very gently for 10 min to avoid loosening the smears. After careful draining, the slide is immersed in 5% (w/v) sodium pyrophosphate solution for 2 min before mounting in buffered glycerol (9 volumes glycerol plus 1 volume sodium carbonate/bicarbonate buffer, pH 9·6), and observing with an oil immersion objective in low-fluorescence oil on a thin cover slip. Far-blue illumination is provided by a quartz-iodine lamp with an FITC interference excitation filter and a BG23 red-absorbing filter (Wild Heerbrugg Ltd, Switzerland). The field is delimited by a 5 mm diameter indexed square graticule with 0·5 mm grid squares.

Counting Procedures

Viable counts

Plates were examined by lateral illumination against a dark background. Conspicuous colonial morphology or pigmentation was noted after 14 d, but details and dominance were recorded after at least another 14 d growth. Counts were made of fast- and slow-growing bacteria, yeasts and filamentous fungi.

Direct counts

Details of the staining and mounting procedure were given on p. 74. Once mounted, the smears were counted as soon as possible to minimize the effects of slow leaching of residual FITC stain making detection of fainter microbes difficult. An oil immersion objective in low-fluorescence oil on a thin cover slip over buffered mountant was used to attain a final magnification of $\times$ 1250. An eyepiece grid-graticule was employed to delimit part of the field of view for counting. Two spots per dilution and six fields per spot were counted, making the procedure similar to those of Clarholm (1973) and Trolldenier (1973).

Gilson Respirometry

Routine screening

Up to 12 core-bearing Quickfit cones were vaselined and fitted into the bases of special flasks (Parkinson and Coups, 1963) which were connected to a Gilson GR14 differential respirometer. An additional 500 ml reference flask was connected to channel 13 (Roger and Dommergues, 1969) to partially compensate for the large total volume of the active flasks. Fluctuations due to residual volume imbalance were minimized by always taking readings at the time of heater activation. 2 ml potassium hydroxide (10 g 100 ml^{-1}) were pipetted into the annular well of each flask to absorb the carbon dioxide evolved, and this was replaced every 10 h. For prolonged incubations the apparatus was flushed with fresh sterile humidified air, which was cooled by passing through 400 cm of thin walled rubber tubing placed in a water bath.

Eight replicate core sections were screened at field sampling temperature before adjusting to the standard temperature (5°C) and equilibrating for 10–15 h before re-screening. Thawed cores were monitored as soon as possible after sampling, but frozen cores were stored below 0°C until microbiological procedures were complete.

Simulations

The Gilson respirometer is also suitable for simulation of environmental changes on peat cores for comparison with observed seasonal fluctuations in microbial components.

Freeze-thaw cycles and temperature changes were simulated in cores *in vitro* using temperature data recorded in the field, thus providing comparative data under controlled conditions. This also eliminated artificial surges in activity provoked by the cutting of moss stems with consequent release of nutrients. Temperatures in the experimental cores were monitored by thermistors.

Attempts were made to restore simulated spring peaks in respiratory activity of cores by adding sterile distilled water or by desiccating them at field temperature to simulate precipitation, meltwater or evaporation and drainage, respectively. The effects of soluble nutrients were tested by adding 0·5 ml of sterile glucose or mannitol solutions (0·5 g 100 ml^{-1}). The effects of nematode predation were studied by adding 0·5 ml suspensions of known nematode density and composition to the core sections.

Re-inoculation of sterilized dried cores with dominant bacterial and fungal strains isolated in the spring bloom is planned to determine the contribution of individual components of the microbial community. Conversely, individual group contributions may be assessed by selective application of antibiotics (Anderson and Domsch, 1974, 1975) to fresh cores in such a system.

In all such experiments parallel flasks containing replicate cores were incubated and similarly treated and subsampled by an 8 mm diam. cork borer to estimate changes in viable and direct count of bacteria, yeasts and filamentous fungi. The holes thus made were plugged with sterile isodiametric Jorgensen tubes to decrease artificial aeration changes.

Decomposition Studies

Enrichment cultures

Cultures were established using peat subsamples added to bacterial and fungal basal media (see pp. 73–74) enriched with cellulose (as homogenized powder, Whatman filter paper or lens tissue), chitin ground to a fine suspension or apple pectin. They were incubated at 10°C for 1 month and then frozen for transport to the UK for subsequent subculturing.

Decomposition of cotton strips

The method of Latter and Howson (1977) is suitable for antarctic field studies. Strips 10 × 35 cm were torn from unbleached calico (43 g m^{-2}, tensile strength 24 kg) with the weft strands lengthwise, and were boiled twice to remove residual starch. After rinsing in distilled water the strips were dried, packed in groups of five between sheets of filter paper wrapped in tinfoil and autoclaved for 15 min at 121°C.

Slits 20 cm deep were made with a knife in the peat profile in the field and 4 cm of the strip were folded round the end of an aluminium strip for vertical insertion. Normally, 6 cm of the strip is left visible at the ground surface, but on Signy Island brown skuas persisted in removing strips so that the ends must either be camouflaged or cut off flush with the moss surface. Ten control strips were inserted and removed immediately to eliminate stretching effects from the measurements.

In each area representative of *Polytrichum*, *Chorisodontium*, *Calliergon* and *Cephaloziella* 20 strips were inserted, 10 to be removed after the spring thaw and 10 after one complete growing season. They were removed by cutting a block of peat away from the face of the strip and peeling it clear using the folded lower portion, and marking the level of the moss surface. After recording and photographing discoloration, the strips were washed in running tap water, dried and stored between filter paper for transport to the UK for measurement of tensile strength.

The fabric used has now been superseded by Shirley Soil Burial Test Fabric (Shirley Institute, available from Dr. D. W. H. Walton, British Antarctic Survey, Cambridge) currently in use on South Georgia (Walton, 1976; Walton and Allsop, 1977).

Environmental Measurements

Water content

The third subsample from each core section was weighed and oven-dried at 70°C to constant weight. Cores from Gilson respirometry experiments were treated similarly after screening.

Loss on ignition

Selected dried subsamples were ignited at 500°C in a muffle furnace to measure total organic content.

pH value

This was measured on field saturated samples directly or on 1:1 or 1:10 dilutions using deionized water.

Bulk density

This was calculated from the wet and dry weight of known volumes of peat from the four depths.

Particulate content

An assessment of mineral and organic particulate content was made by describing and measuring the depth of settled homogenate suspensions in the McCartney bottles.

Aeration

This was assessed simply by inserting triplicate silver-plated brass strips (silver coating 0·05 mm thick on strips measuring 1·2 × 0·3 × 46·0 cm) into each moss stand, and removing one per week to observe the development of darkened areas of silver sulphide indicative of anaerobic conditions.

Microclimate data

Data were collected by an automatic chart recorder (Grant Instruments Ltd, Old Bullback Mill, Barrington, Herts). Single Gulton 32TD25 thermistor probes placed on the surface and at 1·5, 4·5, 7·5 and 10·5 cm below the surface were used at each reference site to provide soil temperature profiles. Incoming radiation was monitored by a Kipp and Zonen CM3 solarimeter on the roof of a nearby field hut. All channels were scanned hourly, and in the UK the charts were calibrated and digitized on a Ferranti Freescan Digitizing Table.

The data were analysed using a PD11/40 computer (Walton, 1977). Peat temperatures at the time of microbiological sampling were recorded using an Edale thermistor thermometer (Grant Instruments Ltd).

Climatic data

Meteorological data were obtained from routine BAS records for Signy Island. Depth of snow and ice or thaw at the sites were measured at the time of sampling.

Vegetation cover

Floristic analyses of all quadrats on SIRS 1 and 2 has been undertaken (R. I. L. Smith, pers. comm.).

Results

Although much statistical analysis remains to be done, the following preliminary results are presented. The range of total peat respiration at field temperatures from 1 cm down to 12 cm was 21–66 ml O_2 uptake m^{-2} h^{-1} at the dry reference site (SIRS 1) and 10–36 ml O_2 uptake m^{-2} h^{-1} at the wet site (SIRS 2). After a sharp increase in respiration rate in spring there was a steep decrease in activity which could be partly eliminated by controlled desiccation and amendation with glucose. However, desiccation was confirmed to be a respiration-limiting factor in *Chorisodontium* peat in late summer.

Preliminary results suggest that nematode predation may be partly responsible for the post-spring decline in microbial population. The seasonal range of bacterial biomass was 1·6–32·6 and 2·1–15·7 mg wet wt. m^{-2} at SIRS 1 and 2, respectively. The corresponding yeast values were occasionally high at SIRS 1, ranging from 7·0 to 558·8 mg wet wt. m^{-2}, but only 2·0 to 4·4 mg wet wt. m^{-2} at SIRS 2. Concomitant counts of fungal colony-generating units were 17·5–76·0 $\times$ 10^8 at SIRS 1 and 3·8–8·5 m^{-2} at SIRS 2. A close correlation between microbial counts and respiratory activity was not apparent. The percentage seasonal fluctuation in biomass of bacteria, yeasts and fungi was 1931%, 7930% and 334%, respectively whereas the corresponding variation in total peat respiration was only 340% at the standard temperature of 5°C.

Simulations have shown an extremely rapid metabolic response of the microbial population to the spring thaw, and the high levels observed, are, to a large extent repeated for each freeze-thaw cycle. The Gilson respirometer has proved eminently suitable for such simulations.

Acknowledgements

The support of the British Antarctic Survey is gratefully acknowledged. The author is indebted by Drs. W. Block and R. I. Lewis-Smith, and J. R. Caldwell, J. C. Ellis-Evans and M. J. Smith for valuable discussions and experimental assistance during these studies, and to Dr. R. M. Laws, Director of BAS for permission to publish this work.

References

ANDERSON, J. P. E. & DOMSCH, K. H. (1974). Use of selective inhibitors in the study of respiratory activities and shifts in bacterial and fungal populations in soil. *Annali de Mikrobiologia Enzimologia* **24,** 189–194.

ANDERSON, J. P. E. & DOMSCH, K. H. (1975). Measurement of bacterial and fungal contributions to respiration of selected agricultural and forest soils. *Canadian Journal of Microbiology* **21,** 314–322.

BABIUK, L. A. & PAUL, E. A. (1970). The use of FITC in the determination of the bacterial biomass of grassland soil. *Canadian Journal of Microbiology* **16,** 57–62.

BAILEY, A. D. & WYNN-WILLIAMS, D. D. Soil microbiological studies at Signy Island, South Orkney Islands. *British Antarctic Survey Bulletin* (in press).

BAKER, J. H. (1970a). Yeasts, moulds and bacteria from an acid peat on Signy Island. In *Antarctic ecology* (Holdgate, M. W., ed.). London and New York: Academic Press, pp. 717–722.

BAKER, J. H. (1970b). Quantitative study of yeasts and bacteria in a Signy Island peat. *British Antarctic Survey Bulletin* **23,** 51–55.

BAKER, J. H. & SMITH, D. G. (1972). The bacteria in an Antarctic peat. *Journal of Applied Bacteriology,* **35,** 589–596.

BEECH, F. W. & DAVENPORT, R. R. (1969). The isolation of non-pathogenic yeasts. In *Isolation Methods for Microbiologists* (Shapton, D. A. & Gould, G. W., eds). Society for Applied Bacteriology Technical Series No. 3. London and New York: Academic Press, pp. 71–83.

CLARHOLM, M. (1973). Direct counts of bacteria in tundra peat for estimating generation time and biomass production. In *Swedish Tundra Biome Project Technical Report* **16** (Flower-Ellis, J. G. K., ed.). Stockholm, pp. 43–56.

COLLINS, N. J., BAKER, J. H. & TILBROOK, P. J. (1975). Signy Island, maritime Antarctic. In *Structure and Function of Tundra Ecosystems* (Rosswall, T. & Heal, O. W., eds). *Ecological Bulletin* No. 20. Stockholm: Swedish Natural Science Research Council, pp. 345–374.

DI MENNA, M. E. (1966). Yeasts in Antarctic soils. *Antonie van Leeuwenhoek* **32,** 29–38.

EKELÖF, E. (1908). Bakeriologische Studien während der Schwedischen Südpolar-Expedition 1901–1903. *Wissenschaftliche Ergebnisse der Schwedischen Südpolar-Expedition 1901–1903,* **4,** 1–120.

HEAL, O. W., BAILEY, A. D. & LATTER, P. M. (1967). Bacteria, fungi and protozoa in Signy Island soils compared with those from a temperate moorland. In *A Discussion on the Terrestrial Antarctic Ecosystem.* Smith, J. E., Organizer. *Philosophical Transactions of the Royal Society, Series B* **252,** 191–197.

LATTER, P. M. & HEAL, O. W. (1971). A preliminary study of the growth of fungi and bacteria from temperate and Antarctic soils in relation to temperature. *Soil Biology and Biochemistry,* **3,** 365–379.

LATTER, P. M. & HOWSON, G. (1977). The use of cotton strips to indicate cellulose decomposition in the field. *Pedobiologia* **17,** 145–155.

MICHELBACHER, A. E. (1939). Seasonal variation in the distribution of two species of *Symphyla* found in California. *Journal of Economic Entomology* **32,** 53–57.

PARKINSON, D. & COUPS, E. (1963). Microbial activity in a podsol. In *Soil Organisms* (Doeksen, J. & van der Drift, J., eds), pp. 167–175. Amsterdam: North Holland Publishing Company.

PARKINSON, D., GRAY, T. R. G. & WILLIAMS, S. T. (1971). *Methods for Studying the Ecology of Soil Micro-organisms*. IBP Handbook 19. Oxford: Blackwell Scientific Publications.

Pirie, J. H. H. (1912). Notes on Antarctic bacteriology. *Report of Scientific Results of the S.Y. Scotia* **3,** 137–148.

Roger, P. & Dommergues, Y. (1969), Utilisation du microrespiromètre differentielle Gilson pour l'étude d'échantillons de sol. *Revue Ecologique et Biologique du sol*, **6,** 299–313.

Rosswall, T. (1971). List of media used in microbiological studies. In *Swedish Tundra Biome Project Technical Report* **8**. Stockholm: Swedish Natural Science Research Council.

Rosswall, T. & Heal, O. W. (eds) (1975). Structure and function of tundra ecosystems. *Ecological Bulletin* 20. Stockholm: Swedish Natural Science Research Council.

Tilbrook, P. J. (1973). The Signy Island Terrestrial Reference Sites. I. An introduction. *British Antarctic Survey Bulletin* **33** and **34,** 65–76.

Trolldenier, G. (1973). The use of fluorescence microscopy for counting soil micro-organisms. In *Modern Methods in the Study of Microbial Ecology* (Rosswall, T., ed.). *Ecological Bulletin* 17. Stockholm: Swedish Natural Science Research Council, pp. 53–60.

Tsiklinsky, M. (1908). La flore microbienne dans les régions du Pole Sud. In *Expedition Antarctique Francaise 1903–1905*. No. 3. Paris: Masson et Cie, 33 pp.

Walton, D. W. H. (1976). 'Shirley test cloth'—a new test cloth for soil burial trials and other studies on cellulose decomposition. *Freshwater Biology* **6,** 299.

Walton, D. W. H. (1977). Radiation and soil temperatures 1972–1974, Signy Island Terrestrial Reference Site. *British antarctic survey data* No. 1. Cambridge.

Walton, D. W. H. & Allsopp, D. (1977). A new test cloth for soil burial trials and other studies on cellulose decomposition. *International Biodeterioration Bulletin* **13,** 112–115.

Williams, S. T. & Gray, T. R. G. (1973). General principles and problems of soil sampling. In *Sampling—microbiological monitoring of environments* (Board, R. G. & Lovelock, D. W., eds). Society for Applied Bacteriology Technical Series No. 7. London and New York: Academic Press, pp. 66–110.

The Spoilage of Vacuum-packed Beef by Cold Tolerant Bacteria

R. H. DAINTY, B. G. SHAW, CHARMAIGNE D. HARDING AND SILVIA MICHANIE*

ARC Meat Research Institute, Langford, Bristol, Avon, England

Introduction

Meat is stored under chill conditions to extend its shelf life and to prevent the growth of food poisoning microbes. Even so after about 3 weeks at 0°C a sweet or sometimes putrid type of spoilage occurs on beef with a normal pH (5·4–5·8) which is unpackaged or wrapped in materials freely permeable to atmospheric gases. This is thought to result from the growth of pseudomonads, which dominate the spoilage flora (Ayres, 1960). The shelf life may be prolonged to 8 weeks or more by vacuum packing in plastic materials with a low permeability to gases, e.g. the permeability to O_2 of commonly used materials ranges from 5–90 cm^3 m^{-2} d^{-1} atm^{-1} O_2 and for CO_2 from 20–300 cm^3 m^{-2} d^{-1} atm^{-1} CO_2. The odours detected on first opening packs which have been stored for a long time are usually described as sour/acid/cheesy, but they are far less objectionable than those of unpackaged meat and often dissipate quickly. The causative organisms are assumed to be lactobacilli which usually dominate the flora (Jaye *et al.*, 1962). If meat of a high pH ($>$ 6·0) is vacuum packaged, spoilage is more rapid and far more offensive. The odours are invariably of a putrid nature (Bem *et al.*, 1976; Taylor and Shaw, 1977) and hydrogen sulphide is produced (Nicol *et al.*, 1970). In view of the common usage of vacuum packaging in both international and home trade, it is important to understand these observations by obtaining as full a knowledge as possible of the microbiological and chemical changes which occur within the pack.

The low permeability of the plastic materials effectively prevents the replacement of oxygen consumed by bacterial and meat tissue respira-

*Present address: S. A. Frigorifico Monte Grande, San Martin 432/525, Monte Grande, Buenos Aires, Argentina.

tion and results in a build-up of the carbon dioxide produced. It is well established that low oxygen tension and/or high carbon dioxide tension can influence the growth of most bacteria so it is necessary to measure the concentrations of these gases in the packs to determine their possible role in determining the nature of microbial growth. The volume of gas inside a typical pack is very small and can be most conveniently removed for analysis by gas chromatography by injecting a quantity of the carrier gas into the pack before sampling. This mixes the gas contents and the measured concentrations are therefore an average for the whole pack. It is not possible to show the variation which might occur between pockets of gas dispersed throughout the pack and this limitation must be borne in mind when interpreting the data.

Microbiological analysis is made of the meat surface because this is where bacterial growth occurs. The dominant types of bacteria are determined by identifying isolates from total viable count plates. Selective media are used to enumerate lactic acid bacteria, Gram negative bacteria and *Brochothrix thermosphacta* so that their numbers can be determined when they are not detected amongst the dominant flora. Hydrogen sulphide-producing bacteria are selectively isolated and identified in experiments with high pH beef on which they are known to be a problem.

The descriptive terms used for the spoilage odours may provide a guide to the chemicals which should be sought in the spoiled meat. Terms such as acid/sour/cheesy suggest the presence of short chain fatty acids, which are common end products of the metabolism of carbohydrates and amino acids by bacteria (Wood, 1961), while sulphury-putrid odours suggest that volatile sulphur compounds and low molecular weight amines are being produced from amino acid breakdown. Substantial concentrations of carbohydrate and amino acids are present in meat and they have been shown to be primary growth substrates for a number of meat spoilage bacteria (Gill, 1976; Newton and Gill, 1978).

Results from two experiments are described below. In the first, normal pH beef was stored in vacuum packs at 0 to 1°C for up to 8 weeks, packs being removed at weekly or fortnightly intervals to determine the pattern of microbial growth. In the second, packs of normal and high pH beef were stored under the same conditions for 9 weeks when they were all removed for microbiological and chemical analysis. Samples were not analysed at intervals because the amount of chemical change in the presence of low microbial numbers is very small and difficult to detect. Representative organisms from vacuum-packed meat were then tested for their ability to produce, in pure culture, compounds detected in the spoiled meat samples.

Methods

Storage of meat

In the first experiment, beef primal joints (chucks), each 3 kg, were obtained, 3 d *post mortem*, from a local slaughterhouse. The joints were vacuum packed in Cryovac BB1 (W. R. Grace Ltd, Elvedon Road, Park Royal, London) and stored at 0–1°C. Two were removed for microbiological examination and gas analysis after 1, 2, 4, 6 and 8 weeks.

Three striploins and three topside joints of both high pH (6·2–6·6) and normal pH (5·4–5·5) were used in the second experiment. They were vacuum packed 3 d *post mortem* in one of three materials: Synthene 40 T/3627 and 38 C/3627 (Smith and Nephew Plastics Ltd, English St, Hull), or Cryovac BB 1, stored at 0–1°C, and analysed microbiologically and chemically after 9 weeks.

Analysis of gas in the packs

The method was essentially that described by Taylor (1964). Sampling patches (2 cm diam.) of Silicone Rubber Sealer (Dow Corning Corporation, Midland, Michigan, USA) were stuck directly on to packs made of Synthene but for Cryovac packs a keying adhesive (Evo-Stick Impact) was first applied. The patches were applied 24–28 h before sampling to allow the sealant to cure. They were positioned over a visibly slack area or a pocket of gas in contact with a lean surface to prevent fat blocking the sampling syringe. One hundred cm^3 of helium gas were injected into the pack via the patch and mixed thoroughly with the gases present by kneading the surface. Twenty-five cm^3 of diluted gases were removed immediately and the syringe connected to the sampling system of a gas chromatograph (Gowmac Instrument Company Ltd, Shannon, Ireland) fitted with a thermal conductivity detector and two copper columns: one 0·45 m long × 35 mm outside diam., filled with 60/80-mesh silica gel to determine carbon dioxide; the other, 1·8 m long × 35 mm outside diam., filled with 60/80 mesh Molecular Sieve 5A allowing both nitrogen and oxygen to be determined. The columns were maintained at 110°C and the carrier gas was helium. Samples of the gases were injected on to each column in turn from a sample cell of 0·5 cm^3 capacity (Taylor, 1964) which was first thoroughly flushed with the gas mixture to ensure there was no contamination of sample. The output from the detector was recorded on a chart strip and the composition of the gases calculated by comparing the relevant peak heights with those obtained from gaseous mixtures of known composition.

Samples for chemical and microbiological analyses were taken immediately after the gas sampling and after a careful note had been taken of the odours detected on first opening the pack.

pH measurement

The pH of the main muscles in each joint was measured before storage, using a Radiometer pH meter Model 29 (Radiometer, Copenhagen, NY Denmark) with a combined electrode (GK 2321C) inserted 2–5 cm into the muscle.

Microbiological analysis

Sampling

In the first experiment lean surfaces were sampled using the techniques of Haines (1937). A 15 cm² area was removed to a depth of 3 mm using a cork borer and chopped with sterile scissors into 100 ml of a sterile saline diluent containing 0·1% (w/v) peptone in which it was homogenized using a Stomacher (A. J. Seward & Co., UAC House, Blackfriars Road, London). In the second experiment 100 cm² of lean surface was sampled using the template and swab technique of Kitchell *et al.* (1973). The swabs were shaken with 10 ml of diluent.

Enumeration of bacteria

One sixtieth of a millilitre drops of selected decimal dilutions of the meat homogenate or swab suspension were transferred to the surface of a variety of media by means of calibrated dropping pipettes (Astell Laboratory Services, Catford, London: Cat. No. 852) and spread. Plate Count Agar (PCA, Oxoid) containing 1% (w/v) NaCl, incubated at 25°C for 5 d, was used to give the total viable count. Lactic acid bacteria were enumerated on Cavett's (1963) modification of Acetate Agar (AA; Rogosa *et al.*, 1951) incubated at 30°C for 48 h in an anaerobic jar under 95% H_2 + 5% CO_2. *Brochothrix thermosphacta*, formerly known as *Microbacterium thermosphactum* but now reclassified (Sneath and Jones, 1976), was enumerated on Streptomycin Thallous Acetate Actidione Agar (STAA; Gardner, 1966) incubated at 20°C for 48 h.

Gram-negative bacteria were enumerated on MacConkey Agar No. 3 (Oxoid) containing 1% glucose (added before autoclaving) incubated at 25°C for 48 h. Only colonies with a diam. of 1 mm or more were counted, colonies smaller than this being invariably Gram positive. Hydrogen sulphide-producing bacteria were enumerated in the second experiment only by counting colonies exhibiting darkening on Lead Acetate Agar (Difco) containing 0·001% (w/v) cysteine after incubation for 48 h at

20°C in an anaerobic jar containing an initial gas mixture of 0·5 % O_2 + 99·5% N_2 (Nicol *et al.*, 1970).

Identification of isolates

The dominant types of bacteria on the meat were determined by identifying isolates from the total viable count plates. Organisms recovered on the MacConkey Agar containing glucose (Gram negatives) and Lead Acetate Agar (H_2S-producers) were also identified.

Where possible, 30 colonies were picked randomly from the lowest countable dilution. The Gram reaction (using the method of Kopeloff and Beerman, 1922) and morphology were determined after 18–24 h growth at 20°C on Nutrient Agar. Identification of the organisms was then carried out using the keys shown in Fig. 1 (Gram negative bacteria) and Fig. 2 (Gram positive bacteria). The methods used were as follows.

Motility. This was determined by phase-contrast microscopy after 24 h incubation at 20°C in Heart Infusion Broth (Difco).

Catalase. The organisms were tested after growth on Nutrient Agar containing 1% glucose for 48 h at 20°C.

Oxidase. After 48 h growth at 20°C on Nutrient Agar, using the method of Kovacs (1956).

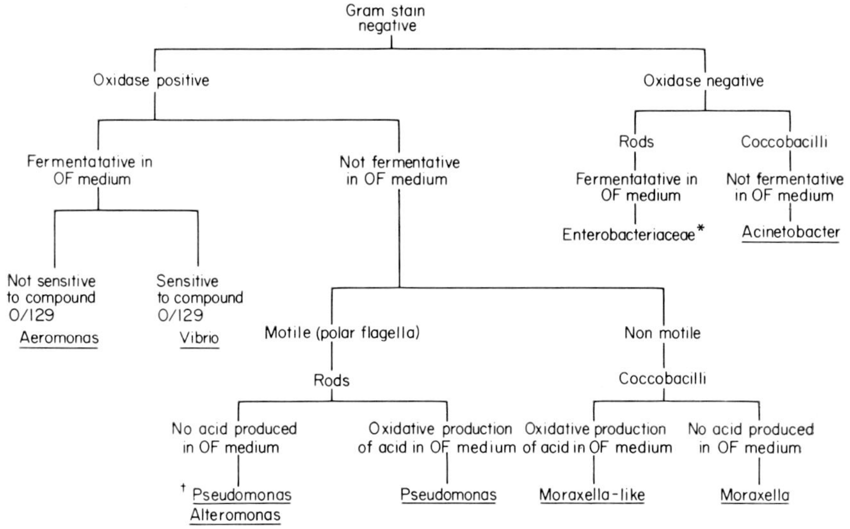

FIG. 1. Identification of Gram negative bacteria. *Enterobacteriaceae can be further identified using the schemes given by Cowan (1974). †*Pseudomonas* spp. and *Alteromonas* spp. can be separated using the scheme given by Lee *et al.* (1977).

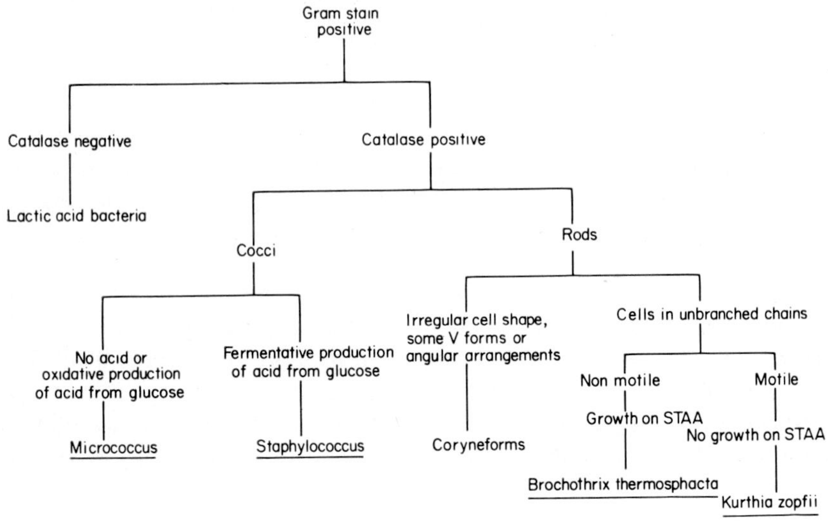

FIG. 2. Identification of Gram-positive bacteria.

Oxidation or fermentation of glucose by Gram negative bacteria. Growth in OF medium (Hugh and Leifson, 1953) was observed daily, up to 7 d, for acid production at 20°C. This medium was also used for detecting gas production which is useful in differentiating *Aeromonas* and *Vibrio* spp.
Oxidation or fermentation of glucose by Gram positive bacteria. The method of the International Sub-Committee on Staphylococci and Micrococci (Anon., 1965) was used.
Sensitivity to vibriostatic agent 0/129. The method of Bain and Shewan (1968) was used with the exception that the 0/129 was dissolved in a 1:1 alcohol/ether mixture.
Growth on STAA. The organisms were streaked on STAA (Gardner, 1966) and observed for growth after 48 h incubation at 20°C.

Chemical analysis

Sampling

Only lean tissue was sampled and to ensure maximum recovery of the spoilage compounds deep tissue was included with the surface layers. Samples were taken as soon as possible after opening the packs and if necessary stored at —20°C until they were extracted.

Extraction and concentration

Amines. The procedure was essentially that described by Patterson and Edwards (1975) for the isolation of amines from pork. One-hundred g samples of defatted meat were finely cut up with scissors into 250 ml of ice cold 5% (w/v) trichloracetic acid (TCA), thus rendering the amines involatile and precipitating soluble meat proteins to avoid excessive frothing during steam distillation. The mixture was macerated in a glass Atomix (MSE Equipment Ltd, Manor Royal, Crawley, Sussex) for 1 min at full speed, centrifuged (5000 × *g*; 15 min) and the supernatant liquid filtered through Whatman No. 1 filters to remove suspended fat and protein. The precipitate was washed by mixing with 100 ml of distilled water followed by centrifugation and filtering as above. The filtrates were combined and transferred to a distillation flask where the pH was adjusted to 10–10·5 by the rapid addition of a calculated volume of 5 M KOH and the flask attached immediately to the all glass distillation train to avoid loss of the very volatile amines. Steam was passed through the alkaline extract at a rate sufficient to produce 3–5 ml min^{-1} of distillate. The outlet from the condenser reached to the bottom of a measuring cylinder, which contained 5–10 ml of 0·1 M HCl, to prevent any loss of the amines. Sixty to seventy ml of distillate were collected, concentrated to dryness in a rotary evaporator and the dry amine hydrochlorides then stored until required for analysis.

Fatty acids. An analogous procedure to that described for amines was used with the following modifications: (a) the meat was extracted with distilled water; (b) for steam distillation the pH was adjusted to between 1·5 and 2·0 with HCl and a few drops of Silicone emulsion FG antifoam (Hopkins and Williams Ltd, P.O. Box 1, Romford, Essex) added to prevent frothing caused by the soluble protein in the extract; (c) 100 ml of distillate were collected with the tip of the condenser outlet beneath the surface of the trapping solution, in this case 5–10 ml of 0·1 M NaOH.

Analysis by gas–liquid chromatography

Amines. The hydrochlorides were dissolved in the minimum volume of distilled water and samples analysed in a Pye 104 instrument (Pye Unicam Ltd, York Street, Cambridge, England) using a 3·3 m long × 6·35 mm outside diam. glass column packed with Chromosorb 103 treated with 15% KOH, except for the first 10 cm which contained Chromosorb 103 (Phase Separations, Deeside Industrial Estate, Queensferry, Clwyd) treated with 25% KOH. The sample was injected directly into the first part of the column where the free amines are regenerated. Injections were made at progressively greater depths to prevent blockage of the column by the KCl formed in the regeneration of the free amines from their hydrochlorides. When the limit of the 25% KOH packing was reached, the top 10 cm was replaced with fresh material. Other conditions were: carrier gas, N_2; flow rate, 30 cm^3 min^{-1}; column and injection port temperature 130°C.

Fatty acids. The sodium salts were dissolved in the minimum volume of water and samples injected on to a column 2·1 m long × 6·35 mm outside diam. packed with 30% (w/w) Carbowax 20 M (Phase Separations) plus 2% (w/w) phosphoric acid on 80/100-mesh Chromosorb W (Phase Separations: acid washed-AW; dimethylchlorosilane treated-DMCS). The carrier gas was argon; flow rate 30 cm^3 min^{-1}; column temperature 120°C; injection port, 200°C.

Both amines and fatty acids were identified initially by comparison of retention times with authentic compounds and the identity then confirmed by combined gas–liquid chromatography/mass spectrometry (LKB 9000: LKB Instruments, Addington Road, Selsdon, S. Croydon). Concentrations were determined by comparison of the peak heights of samples with those of authentic compounds of known concentration.

Results and Discussion

The nature of microbial growth

Changes in total viable counts and counts on the three selective media during storage at 1°C of the vacuum-packed chuck beef joints are shown in Fig. 3. All counts increased during the first 2 weeks after which *B. thermosphacta* stopped growing. Growth of lactic acid bacteria and Gram negative organisms continued between 2 and 4 weeks but subsequently only the lactic acid bacteria grew and they became by far the commonest organisms. Analysis of the flora recovered on PCA containing 1% (w/v) NaCl confirmed the growth pattern shown by the selective media. In general this pattern of growth is typical of that observed by us on many experimental and commercial samples of vacuum packed beef. Slight variations can occur however, e.g. the numbers of Gram negative organisms are often lower than those of *B. thermosphacta* and the lactic acid bacteria sometimes outgrow the latter after, rather than before, 4 weeks.

Identification of the Gram negative bacteria from the chuck joints showed that *Pseudomonas* spp., *Aeromonas* spp. and Enterobacteriaceae all grew in vacuum packs (Table 1). Pseudomonads were the most common Gram negative bacteria after 1 and 2 weeks but Enterobacteriaceae, rarely detected at 1 and 2 weeks, were the most common after 4

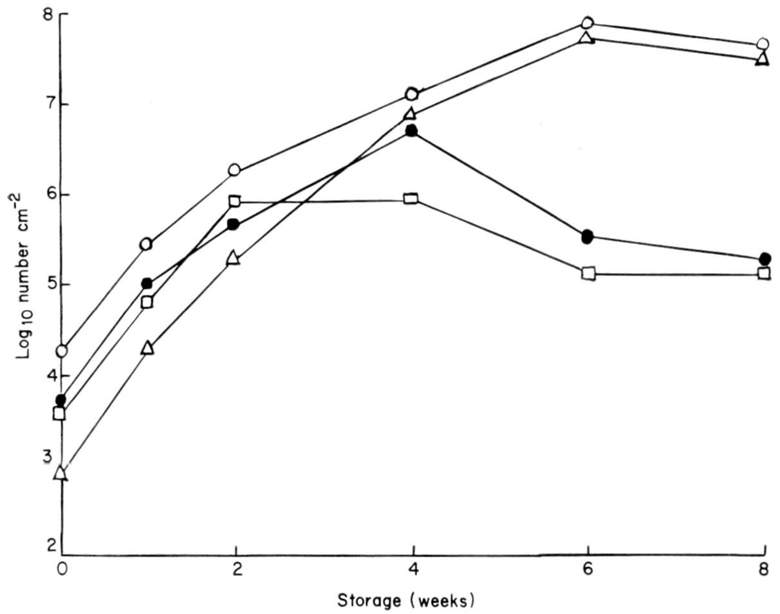

FIG. 3. Bacterial numbers on vacuum packed beef during storage at 1°C. ○, total viable count; △, lactic acid bacteria; □, *B. thermosphacta*; ●, Gram negative bacteria.

weeks. The Enterobacteriaceae were mostly *Serratia liquefaciens*, the rest being *Hafnia* spp. and *Citrobacter*-like organisms.

Other investigators have reported findings which appear to contradict ours. Pierson *et al.* (1970) found that lactic acid bacteria grew to the complete exclusion of pseudomonads, Enterobacteriaceae and *B. thermosphacta*, and Roth and Clark (1972) could not detect growth of pseudomonads or *B. thermosphacta*. These differences may in part be due to differences in the microbial flora on the meat at the time of packing, and, in the case of Gram negative bacteria, may be the consequence of using different selective media.

Reasons for the long shelf life

The extension of shelf life obtained by vacuum packing is primarily a result of the failure of pseudomonads to grow sufficiently to produce spoilage. As Ingram (1962) has suggested, the main reason for this is probably the accumulation of carbon dioxide which has a marked in-

TABLE 1. Numbers of Gram negative bacteria on beef before and after storage at 1°C in vacuum packs

		Log_{10} no. cm^{-2}					
		Pseudomonas spp.		*Aeromonas* spp.		Enterobacteriaceae	
Joint no.	Storage (weeks)	Initial	Stored	Initial	Stored	Initial	Stored
7	1	3·5	4·9	2·9	4·5	2·8	ND
8	2	3·2	5·6	2·3	4·6	2·3	ND
3	4	2·7	ND	2·4	5·8	2·4	6·2

ND, not detected.

hibitory effect on pseudomonads (Haines, 1933). In the experiments on chuck joints, for example, carbon dioxide rose to a concentration of over 20% during the first week of storage, was between 20 and 30% after 2 weeks, and remained between 30 and 40% from 4 weeks onward. In other experiments up to 60% carbon dioxide has been detected. Depletion of oxygen also occurred to concentrations at or below 1% (the minimum we can detect), which may also restrict growth since Clark and Burki (1972) have shown this in laboratory media gassed with 0·5% oxygen.

With the inhibition of pseudomonads, *B. thermosphacta* and *S. liquefaciens* become important as potential spoilage bacteria, and restriction of their growth is necessary to produce a long shelf life. They are probably inhibited to some extent by carbon dioxide (Gardner and Carson, 1967) but this does not explain why they stopped growing after about 3 weeks in our experiments. The lactic acid bacteria growing in the packs are the most likely cause of this. They have been shown to inhibit *B. thermosphacta* by Roth and Clark (1975) and we have recently demonstrated this effect on *S. liquefaciens*. The nature of this inhibition is not fully understood but Roth and Clark (1975) believe it to be caused by a substance produced by the organisms and not by competition for nutrients or by any depression of pH which they might produce.

Spoilage

For a particular pH, the microbiological and chemical findings were essentially the same for the different joints analysed, and therefore only the results for one topside of each pH are given in detail (Tables 2 and 3).

No off-odours were detected on opening the packs of normal pH meat, but a range of fatty acids were found in the meat with acetic and butyric acids being present in by far the greatest concentrations (Table 2). Because a considerable portion of the samples analysed was deep tissue, the figures do not reflect accurately the concentrations of the acids at the surface of the meat, where, since spoilage is a surface phenomenon (Dainty *et al.*, 1975; Gill, 1976), one might expect to find the highest levels. In other samples of meat at normal pH, which smelt sour and cheesy, and in which surface layers constituted a greater proportion of the total sample, we found considerably higher concentrations of acetic (up to 25 000 $\mu g\ kg^{-1}$) and butyric (up to 2000 $\mu g\ kg^{-1}$) acids. The other fatty acids listed in Table 2 were barely detectable. Taking into account the known odours of acetic and butyric acids, it seems reasonable to assume from the data available that these two acids are significant components of the off-odours of this type of meat, with the other acids making little or no contribution. However, there are certain to be com-

TABLE 2. Compounds detected in topsides of beef of different pH values after vacuum packed storage for 9 weeks at 1°C

Compounds	Concentration (μg $100g^{-1}$) in meat of	
	normal pH (5·4–5·5)	high pH (6·2–6·6)
Acetate	3000	12 000
Propionate	12	41
n-Butyrate	160	13 800
Isobutyrate	7	92
n-Valerate	Trace	110
Isovalerate	10	410
4-Methylvalerate	—[a]	Trace
Hexanoate	—[a]	Trace
Methylamine	6	8
Dimethylamine	16	[b]
Trimethylamine	93	2500
Hydrogen sulphide	Absent	Present

[a] Not detected by GLC.
[b] Did not separate completely from trimethylamine on GLC column but was only a minor component.

pounds other than fatty acids contributing to the spoilage odours, and until they are identified an assessment of the relative importance of individual components cannot be made.

The high pH samples were putrid and smelled strongly of hydrogen sulphide. High concentrations of trimethylamine were found in them (Table 2), the concentrations being 20–30 × those found in the corresponding normal pH samples (Table 2) and approximately 50 × those reported by Patterson and Edwards (1975) in their study of aerobically stored samples of pork. It is therefore probable that trimethylamine made a significant contribution to the putrid odours. The samples also contained significantly higher concentrations of fatty acids than the normal pH samples (Table 2). The contribution of these compounds to the spoilage odour was judged, by smell, to be minimal but in similar situations where one might envisage less sulphur compounds and amines, acetic, butyric and possibly isobutyric and *n*- and isovaleric acids, could become more significant.

The different chemical and microbiological pictures obtained for normal and high pH topsides (Tables 2 and 3) give some clues as to which organisms might be responsible for the production of particular compounds—if, indeed, they are all of microbial origin. The very high

TABLE 3. Incidence of micro-organisms ($\log_{10}$ no. cm^{-2}) on topsides of different pH values after vacuum packed storage for 9 weeks at 1°C

pH	Total viable count	Lactic acid bacteria	Enterobacteriaceae	*B. thermosphacta*	*Aeromonas* spp.	*Pseudomonas* spp.	H_2S producers
6·2–6·6	6·8	6·9	6·4	6·2	5·6	5·4	6·2
5·4–5·5	5·6	5·6	4·3	4·1	ND	2·6	3·5

ND, not detected.

counts of hydrogen sulphide-producing bacteria on the high pH samples explains much of the nature of the observed spoilage, the causative organisms being mainly *Pseudomonas* spp. and *Aeromonas* spp. Members of these genera are well documented as producers of H_2S and other sulphur compounds e.g. mercaptans and disulphides (Kadota and Ishida, 1972), and it is probable that such compounds contributed to the sulphury odours detected. It is tempting to attribute the high levels of trimethylamine in the high pH meat to the activities of the large numbers of Gram negatives, which commonly produce amines in pure culture (Laycock and Regier, 1971; Bast, 1971).

Our investigations using meat isolates of lactic acid bacteria and *B. thermosphacta* provide possible explanations for the presence of some of the fatty acids and for their relatively high concentrations in high pH meat. Acetate and isovalerate were produced in substantial concentrations in pure cultures of both types of organism growing in the APT medium of Evans and Niven (1951), with glucose or ribose as the main source of carbon, while *B. thermosphacta* also produced isobutyrate.

The amounts of the acids formed increased as the glucose concentration was lowered or if ribose replaced glucose and it is therefore significant that Gill (1976) has shown that the growth of a lactic acid bacterium on meat stored anaerobically is limited by a depletion of glucose in the surface layers. Consequently, one would predict increased fatty acid production after depletion, which is consistent with the appearance of sour-cheesy odours only after prolonged storage. Furthermore, lower glucose concentrations would be expected in high pH meat than in low pH meat, since the former usually results from a reduction in the glycogen content of the musculature of the live animal (Lawrie, 1974). Thus, glucose would become limiting sooner and further growth would have to be at the expense of some other substrate, e.g. ribose, formed from the breakdown of ATP. We have in fact shown that decreases in ribose concentration can accompany microbial growth on normal pH vacuum-packed meat. Finally, pH itself had a marked effect on fatty acid production in pure bacterial cultures, *B. thermosphacta* for example producing twice as much acetate and four times as much isobutyrate and isovalerate at pH 6·5 as at pH 5·5. Similar effects were demonstrated for lactic acid bacteria.

Although acetate is probably derived from glucose or ribose carbon, isobutyrate and isovalerate are almost certainly produced by deamination of the corresponding amino acids, valine and leucine. Nakae and Elliott (1965a, b) have shown this in a number of lactic acid bacteria isolated from dairy products and we have indirect evidence from carbon balance studies that this is so in *B. thermosphacta*. It is also significant that the production of deaminases by bacteria is repressed by glucose and is greater at higher pH values (Gale, 1951).

However, it does not follow that the acids, particularly the non-branched ones, are necessarily of microbial origin. For example, a sour off-condition in fresh beef, which was associated with the presence of high concentrations of propionic, butyric and acetic acids, and developed in the absence of significant microbial growth, was attributed to a metabolic block of meat enzymes (Shank *et al.*, 1962). We also find acetic and butyric acids in all fresh samples of beef when microbial numbers are very small, and therefore an unlikely source of the acids. Finally, we have found very high concentrations of *n*-butyric acid, 1500 mg kg^{-1} meat, in the surface layers and the deep tissues of a primal topside joint of beef after 18 weeks storage at 1°C. The microbial population was dominated by lactic acid bacteria but we have been unable to demonstrate the formation of the acid by either these organisms grown as a mixed culture, or by pure cultures of any of the other meat strains of lactic acid bacteria and *B. thermosphacta* tested so far. Many dairy strains of lactic acid bacteria (Nakae and Elliott 1965a) do produce *n*-butyric acid, and also propionic and *n*-valeric acids, which were also detected in the spoiled meat, and more data are clearly required in order to establish unequivocally the origin of these acids.

In conclusion, we have obtained evidence that the sour/cheesy odours of normal pH vacuum packed beef may in part be attributed to the appearance of short chain fatty acids, while H_2S and trimethylamine are significant components of the putrid odours of high pH meat packed in the same manner. Pure culture studies are consistent with some but not all of the fatty acids and the hydrogen sulphide being of microbial origin, the lactic acid bacteria and to a lesser extent *B. thermosphacta* being the most likely source of the former. However, definite proof awaits the results of experiments in which pure cultures of the organisms are inoculated on to samples of meat and the organoleptic and chemical changes produced on storage compared with those produced in naturally contaminated and sterile meat samples.

Acknowledgements

We gratefully acknowledge the following for technical assistance: Mr. N. F. Down, Mrs. J. B. Latty, Miss C. M. Hibbard and Mr. C. R. Britton. S.M. is grateful for the award of a United Nations Fellowship.

References

ANON. (1965). International Sub Committee on Staphylococci and Micrococci, Recommendations of sub committee. *International Bulletin of Bacterial Nomenclature and Taxonomy* **15,** 109–110.

AYRES, J. C. (1960). The relationship of organisms of the genus *Pseudomonas* to the spoilage of meat, poultry and eggs. *Journal of Applied Bacteriology* **23,** 471–486.

BAIN, N. & SHEWAN, J. M. (1968). Identification of *Aeromonas, Vibrio* and related organisms. In *Identification Methods for Microbiologists Part B* (Gibbs, B. M. & Shapton, D. A., eds). London and New York: Academic Press, pp. 79–84.

BAST, E. (1971). Uber Vorkommen und Enstehung flüchtiger primarer Amine bei Bakterien. *Archive Mikrobiologie* **79,** 7–11.

BEM, Z., HECHELMANN, H. & LEISTNER, L. (1976). The bacteriology of DFD-meat. *Fleischwirtschaft* **56,** 985–987.

CAVETT, J. J. (1963). A diagnostic key for identifying the lactic acid bacteria of vacuum packed bacon. *Journal of Applied Bacteriology* **26,** 453–470.

CLARK, D. S. & BURKI, T. (1972). Oxygen requirements of strains of *Pseudomonas* and *Achromobacter*. *Canadian Journal of Microbiology* **18,** 321–326.

COWAN, S. T. (1974). *Cowan and Steel's Manual for the Identification of Medical Bacteria* 2nd edn. Cambridge: Cambridge University Press.

DAINTY, R. H., SHAW, B. G., DE BOER, K. A. & SCHEPS, E. S. J. (1975). Protein changes caused by bacterial growth on beef. *Journal of Applied Bacteriology* **39,** 73–81.

EVANS, J. B. & NIVEN, C. F. Jr. (1951). Nutrition of the heterofermentative lactobacilli that cause greening of cured meat products. *Journal of Bacteriology* **62,** 599–603.

GALE, E. F. (1951). *The Chemical Activities of Bacteria* 3rd edn. London: University Tutorial Press.

GARDNER, G. A. (1966). A selective medium for the enumeration of *Microbacterium thermosphactum* in meat and meat products. *Journal of Applied Bacteriology* **29,** 455–460.

GARDNER, G. A. & CARSON, A. W. (1967). Relationship between carbon dioxide production and growth of pure strains of bacteria on porcine muscle. *Journal of Applied Bacteriology* **30,** 500–510.

GILL, C. O. (1976). Substrate limitation of bacterial growth at meat surfaces. *Journal of Applied Bacteriology* **41,** 401–410.

HAINES, R. B. (1933). The influence of carbon dioxide on the rate of multiplication of certain bacteria, as judged by viable counts. *Journal of the Society of the Chemical Industry* **52,** 13T.

HAINES, R. B. (1937). *Microbiology in the Preservation of Animal Tissues*. Department of Scientific and Industrial Research Food Investigation Special Report No. 45. London: HMSO.

HUGH, R. & LEIFSON, E. (1953). The taxonomic significance of fermentative versus oxidative metabolism of carbohydrates by various Gram-negative bacteria. *Journal of Bacteriology* **66,** 24–26.

INGRAM, M. (1962). Microbiological principles in prepacking meat. *Journal of Applied Bacteriology* **25,** 259–281.

JAYE, M., KITTAKA, R. S. & ORDAL, Z. J. (1962). The effect of temperature and packaging material on the storage life and bacterial flora of ground beef. *Food Technology* **16,** 95–98.

KADOTA, H. & ISHIDA, Y. (1972). Production of volatile sulphur compounds by micro-organisms. In *Annual Reviews of Microbiology*, Vol. 26 (Clifton, C. E., Raffel, S. & Starr, M. P., eds). Palo Alto, California: Annual Reviews Incorporated, pp. 127–138.

KITCHELL, A. G., INGRAM, G. C. & HUDSON, W. R. (1973). Microbiological sampling in abattoirs. In *Sampling—Microbiological Monitoring of Environments* (Board, R. G. & Lovelock, D. W., eds). London and New York: Academic Press, pp. 43–61.

KOPELOFF, N. & BEERMAN, P. (1922). Modified Gram stains. *Journal of Infectious Diseases* **31,** 480–482.

KOVACS, N. (1956). Identification of *Pseudomonas pyocyanea* by the oxidase reaction. *Nature, London* **178,** 703.

LAWRIE, R. A. (1974). *Meat Science* 2nd edn. Oxford and New York: Pergamon Press.

LAYCOCK, R. A. & REGIER, L. W. (1971). Trimethylamine producing bacteria on haddock (*Melanogrammus aeglefinus*) fillets during refrigerated storage. *Journal of the Fisheries Board of Canada* **28,** 305–309.

LEE, J. V., GIBSON, D. M. & SHEWAN, J. M. (1977). A numerical taxonomic study of some pseudomonas-like marine bacteria. *Journal of General Microbiology* **98,** 439–451.

NAKAE, T. & ELLIOTT, J. A. (1965a). Volatile fatty acids produced by some lactic acid bacteria. I. Factors influencing production of volatile fatty acids from casein hydrolysate. *Journal of Dairy Science* **48,** 287–292.

NAKAE, T. & ELLIOTT, J. A. (1965b). Production of volatile fatty acids by some lactic acid bacteria. II. Selective formation of volatile fatty acids by degradation of amino acids. *Journal of Dairy Science* **48,** 293–299.

NEWTON, K. G. & GILL, C. O. (1978). The development of the anaerobic spoilage flora of meat stored at chill temperatures. *Journal of Applied Bacteriology* **44,** 91–95.

NICOL, D. J., SHAW, H. K. & LEDWARD, D. A. (1970). Hydrogen sulphide production by bacteria and sulfmyoglobin formation in prepacked chilled beef. *Applied Microbiology* **19,** 937–939.

PATTERSON, R. L. S. & EDWARDS, R. A. (1975). Volatile amine production in uncured pork during storage. *Journal of the Science of Food and Agriculture* **26,** 1371–1373.

PIERSON, M. D., COLLINS-THOMPSON, D. L. & ORDAL, Z. J. (1970). Microbiological, sensory and pigment changes of aerobically and anaerobically packaged beef. *Food Technology* **24,** 1171–1175.

ROGOSA, M., MITCHELL, J. A. & WISEMAN, R. F. (1951). A selective medium for the isolation and enumeration of oral and faecal lactobacilli. *Journal of Bacteriology* **62,** 132–133.

ROTH, L. A. & CLARK, D. S. (1972). Studies on the bacterial flora of vacuum packed fresh beef. *Canadian Journal of Microbiology* **18,** 1761–1766.

ROTH, L. A. & CLARK, D. S. (1975). Effect of lactobacilli and carbon dioxide on the growth of *Microbacterium thermosphactum* on fresh beef. *Canadian Journal of Microbiology* **21,** 629–632.

SHANK, J. L., SILLIKER, J. H. & GOESER, P. A. (1962). The development of a nonmicrobial off-condition in fresh meat. *Applied Microbiology* **10,** 240–246.

SNEATH, P. H. A. & JONES, D. (1976). *Brochothrix*, a new genus tentatively placed in the family Lactobacillaceae. *International Journal of Systematic Bacteriology* **26,** 102–104.

TAYLOR, A. A. (1964). Gas exchanges in packaged meats. I. Headspace gas analysis. *Food Processing and Packaging* **33,** 227–229.

TAYLOR, A. A. & SHAW, B. G. (1977). The effect of meat pH and package permeability on putrefaction and greening in vacuum packed beef. *Journal of Food Technology* **12,** 515–521.

WOOD, W. A. (1961). Fermentation of carbohydrates and related compounds. In *The Bacteria Vol II, Metabolism* (Gunsalus, I. C. & Stanier, R. Y., eds). New York and London: Academic Press, pp. 59–149.

Spoilage Organisms of Refrigerated Poultry Meat

ELLA M. BARNES, G. C. MEAD, C. S. IMPEY AND B. W. ADAMS

ARC Food Research Institute, Colney Lane, Norwich, Norfolk, England

Introduction

The psychrophilic (or psychrotrophic) organisms of poultry meat are defined here as those organisms which can grow to colony size on a suitable agar medium when incubated at 1°C for 14 d. Some can grow at −3°C, most cannot grow above 32–34°C, whilst a few types grow at 28°C but not at 30°C (Barnes and Impey, 1968). Amongst the organisms commonly isolated from carcasses are strains of pigmented and non-pigmented *Pseudomonas* spp., *Alteromonas putrefaciens*, *Aeromonas* spp., *Serratia liquefaciens*, *Flavobacterium* spp., *Cytophaga* spp., *Acinetobacter* spp., *Moraxella* spp., *Brochothrix thermosphacta*, coryneform bacteria, atypical lactobacilli, yeasts and moulds (Lahellec *et al.*, 1975; Barnes, 1976). Of the types present, some, particularly the pseudomonads (*Pseudomonas fluorescens*, *Ps. putida*, *Ps. fragi* etc.) are involved in spoilage, whilst others such as *Flavobacterium* spp. are not generally isolated from the stored carcasses.

The origin, distribution and control of these organisms in the processing plant and factors affecting their multiplication on the processed carcass are discussed below.

Methods

Microbiological examination of the poultry carcass

The sampling method generally used is to take 2 g of skin from the area under the wing together with 3 g of skin and cut muscle surfaces from around the vent (1 g skin = *ca.* 10 cm^2). The bulk sample is macerated in an MSE Ato-Mix blender or a Colworth stomacher with 45 ml of diluent (0·1% peptone, 0·5% sodium chloride). Alternatively, a 5 g sample of neck skin can be removed without seriously damaging the carcass and treated in a similar manner. Ten-fold dilutions are prepared in the same diluent and drops of 0·03 ml are spread over one-quarter of a petri-dish containing the required agar medium, the surface of which had been dried previously. For psychrophilic organisms it is preferable to use a surface-counting method rather than pour plates, since some of these organisms are destroyed at the temperature of molten agar.

Colony count at 1°C

Either Difco Heart Infusion (HI) agar or Oxoid Plate Count Agar (PCA) is used, the plates being incubated at 1°C for 14 d.

Pseudomonads

The medium used to isolate both pigmented and non-pigmented pseudomonads comprises Difco HI Agar supplemented with 50 $\mu g\ ml^{-1}$ cephaloridine, 10 $\mu g\ ml^{-1}$ fucidin and 10 $\mu g\ ml^{-1}$ cetrimide (Mead and Adams, 1977). Plates are incubated at 25°C for 48 h.

If required, plates can be flooded with oxidase reagent (Kovacs, 1956) in order to confirm the presence of *Pseudomonas* spp.

Yeasts and moulds

For the isolation of those yeasts and moulds which can grow at chill temperatures, Bacto Yeast Malt Agar (Difco) at pH 3·4 is used, the plates being incubated for up to 28 days at 1°C.

Isolation and maintenance of psychrophilic bacteria

Colonies from the highest dilution showing growth are inoculated into Difco HI Broth which is incubated at 1°C until turbid. Each culture is then streaked on HI agar to obtain single colonies, the plates being incubated at 20°C for 2–3 d. Representative colonies are picked into HI broth and incubated at 20°C for 24 h. The broth is used to make a stock culture and for a number of tests described below. The organisms may rapidly lose viability, even on the HI agar slopes stored at 1 or 5°C, and representative strains are freeze-dried as soon as possible.

Preliminary identification tests

The HI agar plates are used to record colonial appearance and pigment production and to test for catalase and oxidase (Kovacs, 1956). The 24 h HI broth cultures are examined microscopically for morphology, motility and the Gram reaction. The organisms are then divided into Gram negative and Gram positive groups.

Gram negative organisms

These are non-sporing rods. The 24 h broth culture is used to inoculate (by stabbing) (a) two 1 oz bottles of Hugh and Leifson's medium containing glucose (Hugh and Leifson, 1953), and (b) the arginine medium of Thornley (1960). This and one bottle of the glucose medium are sealed with vaseline. Changes in pH in these media are noted during incubation at 20°C for 14 d. Oxidative or fermentative reactions and gas production are determined from behaviour in the glucose medium, whilst the production of alkaline conditions in the arginine medium indicates arginine dihydrolase activity.

A loopful of the broth culture is streaked across a plate of King *et al.* (1954) agar to test for fluorescin production, several strains being tested on one plate. After overnight incubation at 20°C, the plates are examined under u.v. light to detect fluorescence.

With the above combination of tests it is possible to differentiate strains of *Pseudomonas* spp., *Acinetobacter* spp., *Moraxella* spp. and *Alteromonas putrefaciens* from other Gram negative organisms, in particular *Enterobacteriaceae* and *Aeromonas* spp. (Table 1).

The organism previously described as *Ps. putrefaciens* is now known as *Alteromonas putrefaciens* and one of the poultry reference strains (NCIB 10761) was included by Lee *et al.* (1977) in their detailed study of *Alteromonas*.

The enterobacteria most frequently isolated were identified as *Enterobacter* (now *Serratia*) *liquefaciens* with the help of the late Dr. K. P. Carpenter, Central Public Health Laboratory, Colindale Avenue, London. *Hafnia* spp. have also been found.

Flavobacterium and *Cytophaga* strains are differentiated initially from the other groups by their yellow pigmented colonies. Until recently there has been considerable confusion over the identity of organisms within these genera but now Hayes (1977) has characterized a number of chicken and turkey isolates as part of a comprehensive taxonomic study.

Gram positive non-sporing organisms

These are first differentiated into cocci, coccobacilli or rods which further divide into catalase-negative or -positive types. Many of these strains are not readily identified. The catalase-positive rods are usually *Microbacterium thermosphactum* (McLean and Sulzbacher, 1953), recently renamed *Brochothrix thermosphacta* by Sneath and Jones (1976). Many of the catalase-negative, Gram positive coccobacilli isolated are identical with the group 2 atypical lactobacilli of Thornley and Sharpe (1959).

TABLE 1. Differentiation of *Pseudomonas*, *Alteromonas* and other Gram negative psychrophilic bacteria

Organism	Morphology	Motility (flagella)	Fluorescent pigment	Oxidase	Glucose metabolism	Arginine dihydrolase
Pseudomonas spp. pigmented	Rod	+ (Polar)	+	+	Oxidative	+
Pseudomonas spp. non-pigmented	Rod	+ (Polar)	−	+	Oxidative	+
Alteromonas putrefaciens	Rod	+ (Polar and lateral)	−	+	Oxidative or inert	−
Aeromonas spp.	Rod straight or curved	+ (Polar)	−	+	Fermentative	+
Enterobacteriaceae	Rod	+ or − (Peritrichous)	−	−	Fermentative	−
Acinetobacter spp. (Thornley group C[a])	Coccobacilli	−	−	−	Inert	−
Moraxella spp. (Thornley group A[a] *Acinetobacter*)	Coccobacilli	−	−	+	Oxidative	−
Moraxella spp. (Thornley group B[a] *Acinetobacter*)	Coccobacilli	−	−	+	Inert	−

[a] Thornley (1967).

Control in the Processing Plant

The production of oven-ready poultry involves stunning, slaughter, scalding, plucking, evisceration, washing and chilling. In the UK, the majority of poultry carcasses are water-chilled and frozen but about one-fifth of the total production of broiler chicken is air-chilled and sold fresh.

Psychrophilic spoilage organisms are brought into the processing plant on the outside of the live bird, particularly among the feathers where they may occur at *ca.* 10^8 g^{-1} (Barnes, 1960), strains of *Acinetobacter* and *Moraxella* spp. predominating at this stage. The organisms are not found in the intestine so their ultimate incidence on the processed carcass is unrelated to any faecal contamination.

Many of the psychrophiles are destroyed in the scald tank even when the temperature is as low as 50–51°C which is often used for carcasses sold as chilled products. The low scald temperature ensures that the skin surface remains intact in order to deter multiplication of spoilage organisms on the skin itself and also to safeguard the appearance of the carcass.

During subsequent stages of processing, carcasses are recontaminated with psychrophiles and it has been found that the pseudomonads now form a much higher proportion of the microbial population (Lahellec *et al.*, 1972, 1973). Psychrophiles can be found multiplying on wet surfaces in the plant and are sometimes associated with the water supplies used in washing and chilling.

The main purpose of chilling the warm, freshly-eviscerated carcass is to prevent microbial spoilage. Air-chilling may involve either a preliminary period in a blast tunnel followed by storage in a chill room or storing carcasses directly in a chill room. In both cases it is important that the chilled carcasses are held finally as near as possible to 0°C, since the rate of multiplication of any pseudomonads present will depend on the carcass temperature. One source of carcass contamination is contaminated air in the air-chillers where high counts of psychrophilic yeasts and moulds sometimes occur.

Mechanical immersion chilling ensures a rapid cooling of the carcass and, with appropriate control of water-flow and temperature, carcasses are subjected to a washing effect during chilling which removes a large proportion of the organisms present including the psychrophiles (Mead and Thomas, 1973). In order to achieve this benefit, the water overflow from each chill tank must be sufficient to prevent a progressive accumulation of micro-organisms in the system and the water temperature must be low enough to avoid any problem from bacterial growth.

TABLE 2. Chlorine resistance of poultry spoilage pseudomonads and *Escherichia coli*

Organism	Total no. tested	No. of strains inhibited at each concentration of chlorine (mg l^{-1})					
		0·1	0·5	1·0	2·0	3·0	4·0
Pigmented *Pseudomonas* spp.	27	1	4	6	9	4	3
Non-pigmented *Pseudomonas* spp.	37	1	24	7	3	2	0
Alteromonas putrefaciens	2	0	2	0	0	0	0
Escherichia coli	3	3	0	0	0	0	0

There is considerable debate as to whether a wet carcass held chilled will spoil more rapidly than one where care is taken to ensure a dry skin. However, it is evident that if the water supply contains large numbers of psychrophiles whether it is being used for washing or for chilling, these organisms will increase the contamination of carcasses.

Chlorination

For many years, UK producers have superchlorinated plant water supplies at 5–20 mg l^{-1} of free available chlorine as a means of limiting microbial contamination of carcasses during processing. The water supplies used by processors must be of potable quality but in some cases the water contains large numbers of psychrophilic bacteria, particularly pseudomonads, which occur in varying numbers up to several thousands per ml (Mead and Barnes, 1973). These organisms do not pose a health problem but their presence in process water may contribute significantly to carcass contamination and they are considerably more resistant to chlorine than *Escherichia coli* (Mead *et al.*, 1975) as shown in Table 2. Thus, many of the pseudomonads can survive normal waterworks treatment, which is required to eliminate only coliform bacteria and related organisms; hence the need for additional chlorination at the processing plant. A further advantage of using in-plant chlorination is that it prevents multiplication of micro-organisms on working surfaces and equipment which can lead to slime formation.

Monitoring of processing hygiene

Bacteriological monitoring of carcasses and processing equipment is sometimes necessary as a means of improving control of carcass contamination during processing. The psychrophiles occur as only a small proportion of the total microbial population on carcasses and in order to enumerate these organisms specifically, it is necessary to incubate plates at 1°C for 14 d. The requirement for such a long incubation period is a considerable disadvantage for routine analysis and there is a need for more rapid tests for use in quality control laboratories. No simple medium could selectively isolate all the different types of psychrophiles so emphasis has been placed on the enumeration of the pseudomonads associated with poultry meat spoilage. A medium has been developed by Mead and Adams (1977) which enables counts of pseudomonads to be obtained after incubating plates for only 2 d at 25°C (see Methods).

Processed Carcass

The numbers and types of psychrophilic organisms on the carcass immediately after processing may vary considerably but are usually less than 10^4 cm^{-2} and may be as low as 10^2 cm^{-2}. Usually, these organisms form only a small proportion of the total microbial population.

When carcasses are stored under chill conditions, microbial growth first occurs mainly on the cut muscle surfaces and in the feather follicles. Eventually a sticky slime may form across the skin surface, particularly if the outer cuticle has been removed due to the use of a high scald temperature (58–62°C). When the spoilage organisms reach about 10^7–10^8 cm^{-2}, unpleasant "off-odours" develop and the carcass is unacceptable.

Representative strains of spoilage organisms have been tested by Barnes and Melton (1971) for those properties thought to be important in poultry spoilage. The organisms shown in Table 3 were found to possess a wide range of proteolytic, lipolytic and nuclease activities. When the strains were grown in minced leg muscle a variety of off-odours was produced (Table 3). Detailed studies of the volatile compounds produced by the growth of spoilage organisms on chicken meat have been made by Freeman *et al.* (1976).

Factors affecting shelf life

Amongst the factors which may influence the shelf life of the carcass are the numbers and types of psychrophilic spoilage organisms present on the carcass immediately after processing, the storage temperature, the pH value and type of muscle, the type of packaging and the gaseous environment of the carcass.

Numbers and types of organisms present

The types of psychrophilic organisms which are found most frequently on an eviscerated carcass, both initially and after storage at 1–4°C, are shown in Table 4. The relative proportion of the different species found on the spoiling carcass will depend on a number of factors discussed below. Of particular importance are the pigmented and non-pigmented pseudomonads (*Ps. fluorescens*, *Ps. putida*, *Ps. fragi* and related strains). Analysis of experiments over a number of years has shown that the shelf life of chicken carcasses stored at 1°C in the normal oxygen-permeable polyethylene film is related to the initial numbers of pseudomonads present (Fig. 1), whether these form a major or a minor proportion of the total count at 1°C. It is only when conditions are such that the pseudomonads are inhibited or multiply very slowly that other spoilage organisms predominate.

TABLE 3. Extracellular enzymic activity of the spoilage bacteria

	Pseudomonas spp.			*Alteromonas*	*Acinetobacter*
	Pigmented	Pigmented	Non-pigmented	*putrefaciens*	spp.
No. of strains tested	9	13	31	25	31
Casein hydrolysis	+	—	—	+	—
Gelatin hydrolysis	+	—	—	+	+ (3)[a]
Azocoll breakdown	+	—	—	+	—
Deoxyribonuclease	+ (3)[a]	—	—	+ (13)[a]	+ (28)[a]
Ribonuclease	—	—	+ (2)[a]	+ (24)[a]	— (15)[a]
Lipolysis: tributyrin	+	+ (10)[a]	+	+ (23)[a]	+
Chicken fat	+ (3)[a]	+ (1)[a]	+ (16)[a]	—	+ (16)[a]
Egg yolk reaction	+	+ (3)[a]	—	+ (24)[a]	+ (28)[a]
Off-odours[b]	Putrid, sulphurous	Slightly putrid	Sweetish, stale, urine-like	Putrid, faecal	Fishy

[a] ( ) Number of strains giving the reaction shown.
[b] Representative strain grown in minced chicken leg at 1°C.
Data from Barnes and Melton (1971).

Effect of temperature

Other factors being equal the shelf life of a carcass stored under chill-conditions will depend on the storage temperature and the growth rate of spoilage organisms at the particular temperature used. When nearing the limiting temperatures for growth, not only is the doubling time of the organism very much slower but there is a longer lag phase before growth commences. Farrell and Barnes (1964) found that a typical pseudomonad had a doubling time of 12·4 h at 1°C and one of 36·4 h at —2°C. However, some yeasts and moulds can grow even at —7°C on frozen carcasses (Ingram, 1951).

When air-chilled, eviscerated turkey carcasses were stored in groups of 10 at 5°C, 2°C, 0°C and —2°C, Barnes *et al.* (1978) found that they developed off-odours in an average of 7·2 d at 5°C, 13·9 d at 2°C, 22·6 d at 0°C and about 38 d at —2°C. Analysis of the spoilage flora at —2°C showed that whilst pseudomonads (pigmented and non-pigmented) were present at 10^8 cm^{-2} after 35 and 42 d storage, yeasts identified as *Cryptococcus laurentii* var. *laurentii* and *Candida zeylanoides* were also present at 10^7 cm^{-2} and probably accounted for the unusual, fusty off-odours.

Effect of pH value and type of muscle

Both the composition and pH value of chicken breast and leg muscle are different and may affect the behaviour of the spoilage flora. Growth rates at 1°C in HI broth adjusted to different pH values demonstrated that whilst the pseudomonads grew equally well at pH 5·8 or 6·2, *Alteromonas putrefaciens* grew very much more slowly at 5·8, the pH of chicken breast muscle, than at pH 6·2 which is that of leg muscle (Barnes and Impey, 1968). When a mixture of organisms was inoculated into minced leg and breast muscle, the pseudomonads grew equally rapidly in both tissues but *A. putrefaciens* grew much faster in the leg muscle and strains of *Moraxella* and *Acinetobacter* failed to grow in the breast, but grew in the leg muscle. Thus, one might expect to find a different type of flora growing on the various cut muscle surfaces of a spoiling carcass and spoilage may be more rapid in the areas of high pH value.

The type of packaging and gaseous environment

Evidence is slowly accumulating that the effect of wrapping a carcass in a heat shrunk oxygen-impermeable film is to delay the multiplication of the pseudomonads present. Hence, spoilage is then caused by other groups of organisms which may be present initially in much smaller numbers but are able to multiply inside a pack containing a considerably reduced concentration of oxygen but an increased concentration of CO_2. It was shown by Shrimpton and Barnes (1960) that with chicken

TABLE 4. Psychrophiles on chicken carcasses before and after storage at 1–4°C

Organisms	% Incidence: Initial (100–1000 orgs cm^{-2})	% Incidence: After storage to off-odours (*ca.* 10^8 orgs cm^{-2})
Flavobacteria and *Cytophaga* spp.	20	Not found
Acinetobacter/*Moraxella* spp.	50	10
Pseudomonas spp. (pigmented and non-pigmented)	< 10	70–80
Alteromonas putrefaciens	< 1	10
Yeasts	< 10	May be found when shelf life is extended by controlling growth of pseudomonads
Brochothrix thermosphacta, Atypical lactobacilli[a], *Serratia liquefaciens*, *Aeromonas* spp.	< 10	

[a] Thornley and Sharpe (1959).

carcasses stored at 1°C the use of an oxygen-impermeable film (vinylidene chloride-vinyl chloride copolymer) extended the shelf life to 16 d as compared with about 11 d when an oxygen-permeable polyethylene film was used. Analysis of the gas mixture within the packs (Table 5) showed that the concentration of CO_2 had increased to 9–10% in the impermeable film and the spoilage flora developed much more slowly. An analysis of the spoilage flora given in Table 5 shows that whilst the pseudomonads were the main spoilage organisms on the carcasses wrapped in the oxygen-permeable film, *A. putrefaciens* predominated on the carcasses wrapped in impermeable film and unidentified, Gram positive, catalase-negative cocci also formed a significant part of the flora.

Other studies with ducks (unpublished data) and turkeys (Barnes and Shrimpton, 1968) have confirmed that the growth of pseudomonads is delayed when heat-shrunk, oxygen-impermeable films are used, and the spoilage organisms developing may be *Brochothrix thermosphacta*, the atypical lactobacilli of Thornley and Sharpe (1959) or *Serratia liquefaciens*.

It is now believed that the accumulation of CO_2 within the pack is mainly responsible for delaying the growth of pseudomonads on carcasses wrapped in the oxygen-impermeable film. Early work by Coyne (1933) and Haines (1933) showed that 10–20% CO_2 will delay the growth of pseudomonads and also some other spoilage organisms providing the temperature is kept below 4°C and preferably close to 0°C. Ogilvy and Ayres (1951) showed the value of CO_2 in extending the shelf life of poultry carcasses but noted that an atmosphere containing more than 25% CO_2 caused discoloration and a loss of "bloom".

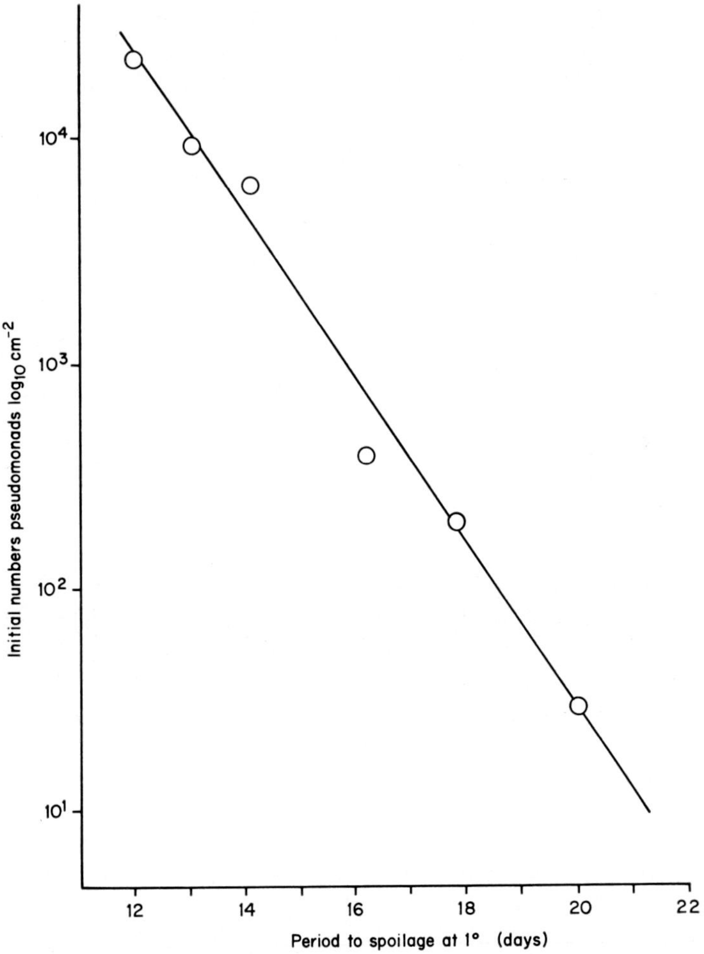

FIG. 1. Relationship between initial numbers of pseudomonads and the shelf life of carcasses held at 1°C (data from 6 studies).

Effect of other treatments

Neither the use of antibiotics such as chlortetracycline nor the use of irradiation is permitted at the present time but in both cases evidence suggests that these treatments delay the multiplication of pseudomonads. When chlortetracycline was used, yeasts together with *Acinetobacter* spp. formed a major part of the spoilage flora of carcasses held at 1°C (Barnes and Shrimpton, 1958). By contrast, *Acinetobacter* spp. and atypical lactobacilli were isolated from stored, irradiated carcasses (Ingram and Thornley, 1959).

TABLE 5. The spoilage flora of chickens stored at 1°C wrapped in oxygen-permeable film (A)[a] and oxygen-impermeable film (B)[b]

Treatment of carcasses and storage time (d)	Composition of gas within the bags		Colony count at 1° (orgs cm^{-2})	No. of strains examined	% Distribution of strains				
	Oxygen (%)	Carbon dioxide (%)			*Pseudomonas* spp. pigmented and non-pigmented	*Alteromonas putrefaciens*	*Acinetobacter* spp.	Unidentified, catalase -ve, Gram + ve cocci	Unclassified strains
Film A[a]									
0	20·8	0·2	$2·7 \times 10^{4}$[c]	22	76	19	5		
8	20·2	0·5	$4·9 \times 10^{6}$	21	97		3		
12	17·1	2·1	$2·8 \times 10^{8}$	28	89	3			8
Film B[b]									
0	15·1	4·4	$2·7 \times 10^{4}$	22	68	9	19		4
8	6·5	6·2	$5·2 \times 10^{5}$	23	43	42	3		12
12	2·8	9·1	$6·6 \times 10^{6}$	26	3	69		28	
14	3·9	9·3	$7·5 \times 10^{6}$	27	4	60		36	
16	6·7	10·0	$5·8 \times 10^{7}$	26	38	51		4	7

[a] Oxygen-permeable film (polyethylene).
[b] Oxygen-impermeable film (vinylidene chloride-vinyl chloride copolymer).
[c] Each count is the geometic mean of results from 3 carcasses.
Data from Shrimpton and Barnes (1960).

Summary

The different types of psychrophilic (or psychrotrophic) organisms associated with poultry carcasses are described. Of these, the pseudomonads are considered to be the most important in relation to spoilage. A selective medium has been developed for the isolation and enumeration of the pseudomonads in order to facilitate monitoring and control of these organisms in the processing plant.

Factors are considered which may affect the type of spoilage and shelf life of the carcass. These include the numbers and types of psychrophilic spoilage organisms present on the carcass immediately after processing, the storage temperature, the pH and type of muscle, the type of packaging and the gaseous environment of the stored carcass.

References

BARNES, E. M. (1960). The sources of the different psychrophilic spoilage organisms on chilled eviscerated poultry. *Proceedings of the 10th International Congress of Refrigeration, Copenhagen* Vol. 3. London: Pergamon Press, pp. 97–100.

BARNES, E. M. (1976) Microbiological problems of poultry at refrigerator temperatures—a review. *Journal of the Science of Food and Agriculture* **27,** 777–782.

BARNES, E. M. & IMPEY, C. S. (1968). Psychrophilic spoilage bacteria of poultry. *Journal of Applied Bacteriology* **31,** 97–107.

BARNES, E. M. & MELTON, W. (1971). Extracellular enzymic activity of poultry spoilage bacteria. *Journal of Applied Bacteriology* **34,** 599–609.

BARNES, E. M. & SHRIMPTON, D. H. (1958). The effect of the tetracycline compounds on the storage life and microbiology of chilled eviscerated poultry. *Journal of Applied Bacteriology* **21,** 313–329.

BARNES, E. M. & SHRIMPTON, D. H. (1968). The effect of processing and marketing procedures on the bacteriological condition and shelf-life of eviscerated turkeys. *British Poultry Science* **9,** 243–251.

BARNES, E. M., IMPEY, C. S., GEESON, J. D. & BUHAGIAR, R. W. M. (1978). The effect of storage temperature on the shelf-life of eviscerated air-chilled turkeys. *British Poultry Science* **19,** 77–84.

COYNE, F. P. (1933). The effect of carbon dioxide on bacterial growth. *Proceedings of the Royal Society, Series B* **113,** 196–216.

FARRELL, A. J. & BARNES, E. M. (1964). The bacteriology of chilling procedures used in poultry processing plants. *British Poultry Science* **5,** 89–95.

FREEMAN, L. R., SILVERMAN, G. J. ANGELINI, P., MERRITT, JR. C. & ESSELEN, W. B. (1976). Volatiles produced by microorganisms isolated from refrigerated chicken at spoilage. *Applied and Environmental Microbiology* **32,** 222–231.

HAINES, R. B. (1933). The influence of carbon dioxide on the rate of multiplication of certain bacteria, as judged by viable counts. *Journal of the Society of Chemical Industry, London* **52,** 13T–17T.

HAYES, P. R. (1977). A taxonomic study of flavobacteria and related Gram negative yellow pigmented rods. *Journal of Applied Bacteriology* **43,** 345–367.

HUGH, R. & LEIFSON, E. (1953). The taxonomic significance of fermentative versus oxidative metabolism of carbohydrates by various Gram-negative bacteria. *Journal of Bacteriology* **66,** 24–26.

INGRAM, M. (1951). The effect of cold on microorganisms in relation to food. *Proceedings of the Society for Applied Bacteriology* **14,** 243–260.

INGRAM, M. & THORNLEY, M. J. (1959). Changes in spoilage pattern of chicken meat as a result of irradiation. *International Journal of Applied Radiation and Isotopes* **6,** 122–128.

KING, H. O., WARD, M. K. & RANEY, D. E. (1954). Two simple media for the demonstration of pyocyanin and fluorescin. *Journal of Laboratory and Clinical Medicine* **44,** 301–305.

KOVACS, N. (1956). Identification of *Pseudomonas pyocyanea* by the oxidase reaction. *Nature, London* **178,** 703.

LAHELLEC, C., MEURIER, C. & CATSARAS, M. (1972). La flore psychrotrophe des carcasses de volailes. I. Evolution aux différents postes d'une chaine d'abbatage. *Annales de Recherches Véterinaire* **3,** 421–434.

LAHELLEC, C., MEURIER, C. & CATSARAS, M. (1973). La flore psychrotrophe des carcasses de volailes. II. Evolution au cours de l'evisceration. *Annales de Recherches Véterinaire* **4,** 499–512.

LAHELLEC, C., MEURIER, C., BENNEJEAN, G. & CATSARAS, M. (1975). A study of 5920 strains of psychrotrophic bacteria isolated from chickens. *Journal of Applied Bacteriology* **38,** 89–97.

LEE, J. V., GIBSON, D. M. & SHEWAN, J. M. (1977). A numerical taxonomic study of some *Pseudomonas*-like marine bacteria. *Journal of General Microbiology* **98,** 439–451.

MCLEAN, R. A. & SULZBACHER, W. L. (1953). *Microbacterium thermosphactum* spec. nov., a non heat resistant bacterium from fresh pork sausages. *Journal of Bacteriology* **65,** 428–433.

MEAD, G. C. & ADAMS, B. W. (1977). A selective medium for the rapid isolation of pseudomonads associated with poultry meat spoilage. *British Poultry Science* **18,** 661–670.

MEAD, G. C. & BARNES, E. M. (1973). Some factors which may affect the bacteriological quality of water-chilled poultry carcasses. *Poultry Meat Symposium, Roskilde, Denmark*, paper A6, pp. 1–16.

MEAD, G. C. & THOMAS, N. L. (1973). The bacteriological condition of eviscerated chickens processed under controlled conditions in a spin-chilling system and sampled by two different methods. *British Poultry Science* **14,** 413–419.

MEAD, G. C., ADAMS, B. W. & PARRY, R. T. (1975). The effectiveness of in-plant chlorination in poultry processing. *British Poultry Science* **16,** 517–526.

OGILVY, W. S. & AYRES, J. C. (1951). Post-mortem changes in stored meats. II. The effect of atmospheres containing carbon dioxide in prolonging the storage life of cut-up chicken. *Food Technology, Champaign* **5,** 97–102.

SNEATH, P. H. A. & JONES, D. (1976). *Brochothrix*: a new genus tentatively placed in the Family *Lactobacillaceae*. *International Journal of Systematic Bacteriology* **26,** 102–104.

SHRIMPTON, D. H. & BARNES, E. M. (1960). A comparison of oxygen-permeable and impermeable wrapping materials for the storage of chilled eviscerated poultry. *Chemistry and Industry* 1492–1493.

THORNLEY, M. J. (1960). The differentiation of *Pseudomonas* from other Gram negative bacteria on the basis of arginine metabolism. *Journal of Applied Bacteriology* **23,** 37–52.

Thornley, M. J. (1967). A taxonomic study of *Acinetobacter* and related genera. *Journal of General Microbiology* **49,** 211–257.

Thornley, M. J. & Sharpe, M. E. (1959). Micro-organisms from chicken meat related to both lactobacilli and aerobic spore formers. *Journal of Applied Bacteriology* **22,** 368–376.

The Microbial Spoilage of Fish with Special Reference to the Role of Psychrophiles

J. M. SHEWAN AND C. K. MURRAY

Torry Research Station, 135 Abbey Road, Aberdeen, Scotland

Introduction

When an animal such as a fish, hen or ox dies, there is immediately set in motion a series of muscle enzymic reactions involving, among others, adenosine triphosphate (ATP), creatine phosphate and glycogen. These result, even within a very short time and certainly long before microbial action has begun, in the production of a large number of low molecular weight compounds (Shibata, 1977) such as inosine, ribose, lactate and creatine which together with the other muscle extractives are most readily attacked by bacteria, leading eventually to the production of spoilage odours and flavours.

However, in fish, these enzymic reactions, together with any muscle proteolysis which may occur, do not result in any markedly undesirable organoleptic changes since sterile pieces of cod can be stored at chill temperatures (0–5°C) for periods of weeks with little or no apparent change in their sensory properties other than minor textural changes. It is only after bacterial attack that spoilage leading to unacceptable organoleptic changes occur.

Bacteriology of Newly Caught Fish

It has already been established that in newly caught fish, the muscle is sterile but the gills, the integument and, in "feedy" fish, the intestines carry 10^2 to 10^5, 10^3 to 10^9 and 10^3 to 10^9 bacteria cm^{-2} of skin or g^{-1} of gill tissue and intestinal content, respectively (Shewan, 1976).

The numbers and types of bacteria present are related to the environment in which the fish are caught and in general fish from tropical waters have more bacteria than those from temperate zones and fresh-water fish have fewer bacteria than marine fish (Shewan, 1976). The types present reflect the flora of the environment from which the fish are caught (Shewan, 1976); temperate zone fish are characterized by a predominance of psychrotrophic or psychrophilic Gram negative bacteria of the genera *Pseudomonas*, *Alteromonas* and *Moraxella*, whereas tropical and subtropical fish have mesophilic Gram positive types, such as coryneforms and micrococci.

Definition of Psychrophile

It would be useful at this point to define the terms psychrophile, psychrotroph and mesophile.

Forster (1887, 1892) was probably the first to record the growth of bacteria at temperatures of 0°C or below, but it was Schmidt-Nielsen (1902) who first proposed the term psychrophile for micro-organisms which not only tolerated, but could actually grow at, low temperatures. Since then there has been considerable confusion over the use of the term because it came to be recognized that some mesophiles although having optimal temperatures of between 25 to 30°C could grow relatively quickly at temperatures in the region of 0°C, whereas other mesophiles failed to grow below 4°C. Moreover, there are some psychrophiles which have optimal temperatures below 20°C and indeed may be killed if kept for any length of time at temperatures of 20°C and above. The latter are considered to be true or obligate psychrophiles, the other facultative cold- or psychro-tolerant or psychrotrophic. These and other temperature relationships are given in Table 1.

Microbial Invasion of Fish Tissues and Associated Sensory Changes during Spoilage

It is generally assumed that once a fish dies and the body defences against microbial invasion cease to operate, bacteria would readily invade the muscle tissue by way of the vascular tissue from the gills along the caudal vein, through the skin and particularly through the "pores" of the lateral line and through the intestinal tract. If this is so, it should be comparatively easy to demonstrate histologically or by bacterial culture the presence of bacteria in the muscle tissue.

It has been shown that the muscle of freshly caught fish, such as cod, is sterile and during the first few days of storage in ice very few bacteria can be demonstrated by histological examination with the light microscope. Even by scanning electron microscopy only a few isolated microbial cells can be seen (Fig. 1) on the epidermis. If the whole cod is allowed to remain dry under conditions of low r.h. at 0°C the surface micro-organisms

TABLE 1. Temperature relationships of micro-organisms

Minimum growth temperature (°C)	Optimal growth temperature (°C)	Maximum growth temperature (°C)	Terminology
−7 to −10	12–15	*ca.* 18	Obligate psychrophile *or* cryophile
ca. 0	20–30	37–40	Facultative psychrophile *or* psychrotroph *or* cold tolerant
5 to 7	37	40–45	Mesophile
Just above 37	55		Obligate thermophile Stenothermophile
37	55		Facultative thermophile Eury-thermophile

do not appear to flourish except in localized areas containing trapped moisture. The gills naturally offer such ideal area for growth.

Using histological methods, organisms in the fish gut can be observed penetrating the peritoneal lining after about 72 h. Under the dry conditions, it takes from 7 to 10 d before significant organoleptic changes can be discerned in the tissue, near the head and belly regions. Histologically, the surface of the whole fish still shows little evidence of bacterial activity and only after 21 d or so can significant numbers of organisms be demonstrated in the tissue, mainly near the head and belly region. Immediately beneath the skin there are only a small number of micro-organisms which appear to have gained entry to the tissue through the pores along the lateral line.

The histological picture in whole iced cod, stored at ambient 2·5°C, is quite different because moist conditions prevail over the entire surface of the fish. Washing the fish surface by ice melt water, although effective in reducing the bacterial population in the early stages of storage, has little or no effect on the surface numbers as the load of psychrophiles increases. Even after 12–14 d of storage, although a large number of micro-organisms can be demonstrated on the fish surfaces (Fig. 2), and organoleptic changes are now discernible in the fish fillet, microscopical fields only occasionally contain organisms (1 bacteria field^{-1} at a magnification of 500 × is equivalent to 10^6 bacteria g^{-1}). However, if the temperature of storage is raised to 8°C (i.e. the fish temperature is at 8°C) a different picture again emerges. Histological examination of muscle blocks taken from whole cod stored at 8°C for 12 d under dry conditions showed large numbers of bacteria invading the tissue via the collagen fibres, particularly the epimysium (Fig. 3), although the numbers of surface bacteria were still comparatively few. As might be expected, strong spoilage flavours were detected earlier in fillets taken from such fish and

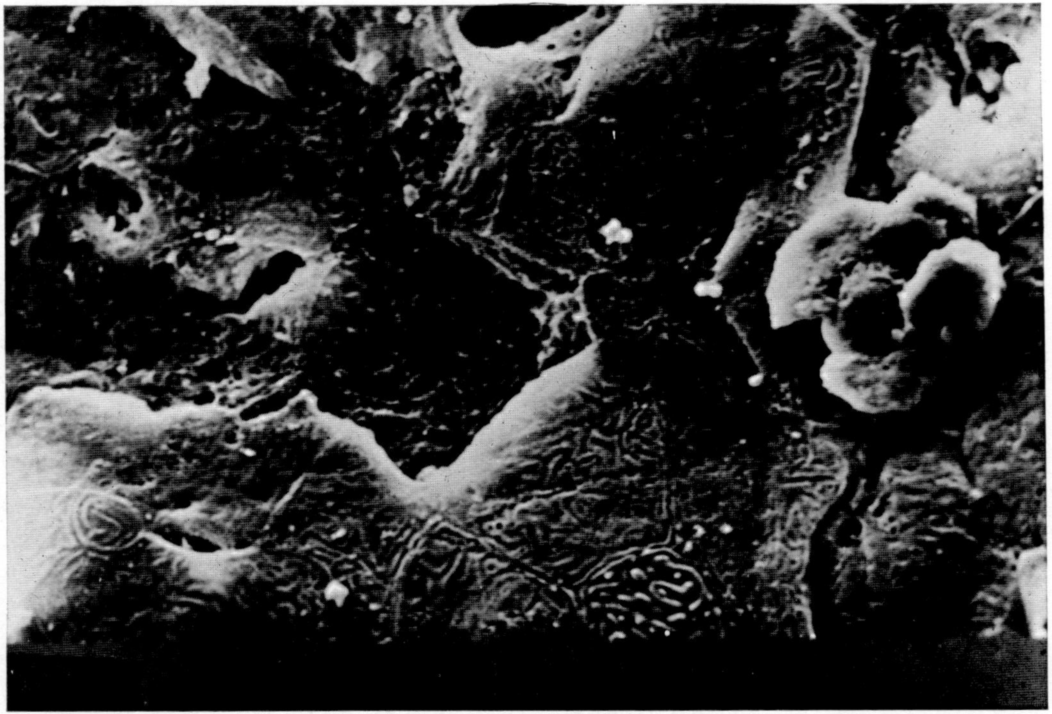

Fig. 1. Scanning electron microscope of fresh cod epidermis.

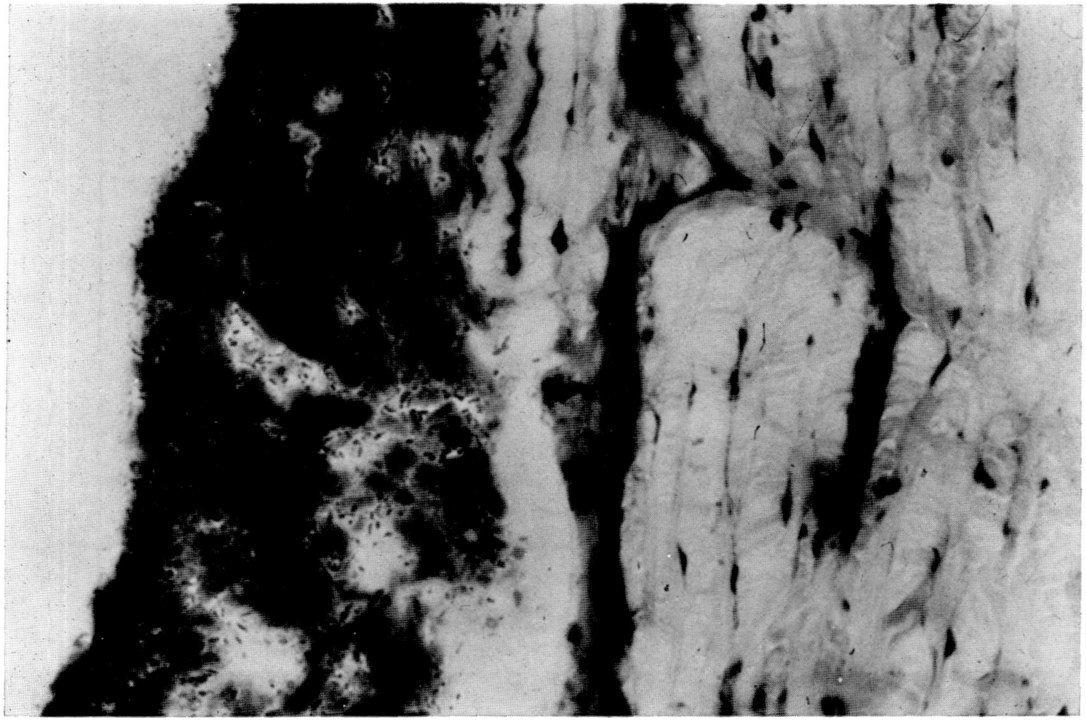

FIG. 2. Cod skin, stored 14 d at 2·5°C in ice (Giemsa) × 480.

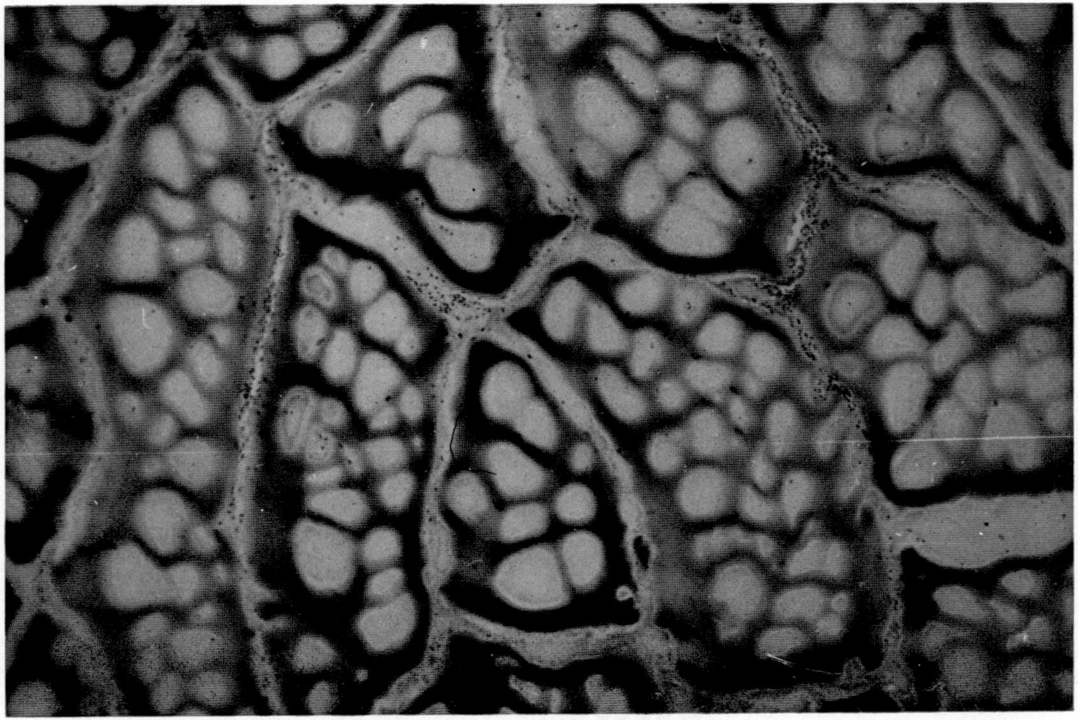

FIG. 3. Cod muscle, stored 12 d at 8°C (Giemsa) × 300.

the deterioration of the gut and peritoneum was much more rapid as a consequence of the marked penetration of the bacteria.

It may be concluded, therefore, that when fish are stored at chill temperatures, the organoleptic and chemical processes leading eventually to the rejection of the fish as food is mainly a bacteriological process confined mostly to the integument and slime resulting from the attack on the low molecular weight compounds present in the skin and slime or diffusing outwards from the flesh. On the other hand, at higher temperatures of storage, actual bacterial penetration of the flesh occurs with consequential attack on the muscle constituents.

So far it has not been possible by direct methods to identify even to the genus level the bacterial types causing invasion of the flesh at temperatures above 5°C so it is not known whether the spoilage bacteria below 5°C and 7·5°C are the same.

Changes in the Microbial Flora during Spoilage

It has already been stated that in newly caught fish the skin floras may differ considerably in their composition depending on the environment from which they have been caught (Shewan, 1976). It is somewhat surprising to find, therefore, that after storage at chill temperatures or in ice for periods of 7 d or more (when bacterial spoilage is detectable, both organoleptically and by chemical methods) the floras are all dominated by Gram negative bacteria particularly psychrophilic or psychrotrophic types of the genera *Pseudomonas*, *Alteromonas* and *Moraxella* (Table 2). One of the main reasons for this is undoubtedly the fairly rapid growth of these organisms at chill temperatures outgrowing the mesophilic Gram positive types. As Table 3 shows, most of the species known to cause spoilage of fish such as *Ps. fluorescens*, *Ps. fragi*, *A. putrefaciens*, *Moraxella* sp. can grow at temperatures of —5 to —6°C and similar species have generation times of from 10 to 30 h at 0°C (Table 4). Consequently it is easy to see why, although present in small numbers in the newly caught fish, they increase to 10^5 to 10^7 g^{-1} after 7 d at chill temperatures.

Many other proteinaceous foods such as meat, poultry, crustacea and milk, when stored at chill temperatures, develop floras, consisting mainly of pseudomonads (Table 5), irrespective of their differing initial floras and despite the wide differences in the chemical composition of their extractives (Table 6). The organoleptic changes induced in sterile cod muscle by the action of pure cultures of such organisms are shown in Table 7. Experience has indicated that the volatile sulphur compounds such as $H_2S(CH_3)_2S$ and CH_3SH are formed during the spoilage of fish (Herbert *et al.*, 1975; Herbert and Shewan, 1975), and it is known that

TABLE 2. Changes in the bacterial floras of fish after storage at chill temperatures expressed as a percentage of isolates examined

	Storage (d)							
Bacterial types	0[a]	7[a]	0[b]	8[b]	Fresh[c]	Spoiled[c]	0[d]	10[b]
Pseudomonas sp. 1			14	3				
Pseudomonas sp. 2	20	53	14	53	18	57	14	50
Pseudomonas sp. 3			3	31				
Achromobacter sp.	25	13	—	—	—	—	33	38
Acinetobacter sp.	—	—	32	8	1	4	—	—
Moraxella sp.	—	—	19	3	8	26	—	—
Flavobacterium sp.	25	1	18	0	8	—	4	0
Coryneforms	16	2	—	—	12	1	41	12
Micrococcus	1	—	—	—	49	7	8	0
Vibrio sp.	13	31	—	—	1	—	—	—
Others	—	—	—	2	3	5	—	—

[a] Simidu *et al*. (1969).
[b] Lee and Harrison (1968).
[c] Gillespie and Macrae (1975).
[d] Shewan *et al*. (1960).

TABLE 3. Minimum growth temperatures of various species of psychrophiles[a]

Bacterial strain	Origin of species	Temperature in °C (after 14 d observation)
Vibrio anguillarum (NCMB 6)	Ulcerative lesion in plaice	—1
Pseudomonas fluorescens (NCMB 129)	Freshly caught cod	—4
Ps. fragi (NCMB 320)	Cod, in ice for 2 d	—6·5
Ps. putida (NCMB 406)	Freshly caught cod	—4·0
Pseudomonas sp. (NCMB 1520)	12 d iced cod	—4·0
Ps. putrefaciens (NCMB 1735)	Unknown	—6·0
Ps. putrefaciens (NCIB 8615)	Surface tainted butter	—2·0
Ps. rubescens (NCIB 8767)	Cutting oil	—3·0
Moraxella-like species 78	Spoiling prepacked cod fillets	—5·0

[a] After Harrison-Church (1973).

even in low concentrations these compounds can contribute significantly to the organoleptic properties of spoiling fish (Table 8).

Differing Rates of Spoilage of Various Fish Species

If it be accepted that the spoilage of fish stored at chill temperatures or in ice occurs at or in the surface slime, and the organisms therein are dependent on environment, then it would be logical to conclude that the spoilage of any fish caught in the same environment and stored under identical conditions would spoil at the same rates. However, various species of fish, even those belonging to the same family such as the gadoids (whiting, cod, coalfish), spoil at different rates. It has, of course, been well known to the practical fisherman for decades that whiting spoil very rapidly, much more so than codling of the same size, and considerably quicker than plaice and other flatfish. These different spoilage rates can be expressed by organoleptic assessment and by chemical or physical measurements; both show that the spoilage rates of different species of fish differ remarkably (Table 9). When such studies are extended to tropical species stored in ice, even more remarkable periods of keeping fresh have been claimed (Shewan, 1976). Table 9, for example, shows that Indian bream is considered still edible after storage for up to

TABLE 4. Reported generation times for some psychrophiles at 0 to 20°C[a]

		Generation time (h) at				
Bacterial strain	Source	0°C	2–2·5°C	4–5°C	10°C	20°C
Cl. hastiforme	Milk	—	—	73·0	25·5	—
Clostridium sp. (61)	River mud	17	—	—	9	3·5
Bacillus W 25	Soil mud	23	—	8·5	6	2·5
Pseudomonas 92	Dairy product	26·6	—	11·7	5·4	1·7
Ps. fluorescens	Fish	30·2	—	6·7	—	1·4
Ps. fragi	Dairy product	11·3	7·7	5·0	2·6	1·1
Pseudomonas sp. (1–3b)	Chicken	10·3	8	7	2·7	1·6
Gram negative rod	Fish	20·0	—	7·6 (6°)	1·9 (12°)	—
Pseudomonas sp. (451)	Meat		13·8	9·7	4·0	1·2
Ps. fluorescens		30·20				
Ps. fluorescens		26·4		10·65		
Pseudomonas sp.		10·33			2·66	
Ps. fluorescens				4·2 to 8·20		

[a] Adapted from Tompkin (1973) and Morita (1975).

TABLE 5. Major differences in the muscle extractives (from various sources) in mg 100 g^{-1} wet wt. of material[a]

	Cod	Herring	Dogfish	Lobster	Poultry leg muscle	Mammalian muscle
Total extractives	1200	1200	3000	5500	1200	3500
Total free amino acids	75	300	100	3000	440	350
Arginine	<10	<10	<10	750	<20	<10
Glycine	20	20	20	100–1000	<20	<10
Glutamic acid	<10	<10	<10	270	55	36
Histidine	<1·0	86	<1·0	—	<10	<10
Proline	<1·0	<1·0	<1·0	750	<10	<10
Creatine	400	400	300	Absent	—	550
Betaine	Absent	Absent	150	100	—	—
Trimethylamine oxide	350	250	500–1000	100	0	0
Anserine	150	Absent	Absent	Absent	280	150
Carnosine	Absent	Absent	Absent	Absent	180	200
Urea	Absent	Absent	2000	—	—	35

[a] After Shewan (1974).

TABLE 6. Changes in the bacterial floras of various foods during chill storage expressed as a percentage of the bacterial isolates

Microbial group	Food product: Milk and milk[a] product	Chicken[b]		Fish[c]		Meat[d]		Shellfish[e]		Ground beef	
	7 d[g] at 4°C	Initial flora	10–11 d at 1°C	Initial flora	10 d iced	Initial flora	4 d at 7°C	Initial flora	10 d iced	Initial flora	14 d at 7°C
Pseudomonas sp.											
Pigmented	45	2	51	2	5	9		8	2		
Non-pigmented	25	—	20	12	45	11	85	17	80	11	84
Ps. putrefaciens	—	—	10								
Achromobacter sp.	10	7	7	33	38	27		24	13		
Alcaligenes sp.	10	—	—	—	—	—	—	—	—	—	—
Flavobacterium sp.	1	14		4	0			17	2	0	0
Micrococcus sp.		50		8	0	53		2	2	50	5
Coliforms	9	8									
Gram positive rods		14		41	12			26	0		
Others		5	12	0			15	6	1	14	11
Yeasts										25	0

[a] Schultze and Olsen (1960). [b] Barnes and Thornley (1966). [c] Shewan *et al.* (1960).
[d] Ienistea *et al.* (1970). [e] Hobbs *et al.* (1971). [f] Rogers and McCleskey (1957).
[g] No data on the original flora.

TABLE 7. Organoleptic changes produced in sterile fish muscle stored at 0 to 2°C by pure cultures of bacteria

	Storage period		Possible chemical components
	ca. 7 d	14 d	
Pseudomonas fluorescens			
NCMB 129 (from fish)	Slightly sulphidy with NH_3	Strong NH_3, amines sour sink	NH_3, TMA, H_2S, $(CH_3)_2S$ $CH_3{\cdot}SH$
NCMB 1792 (from fish)		Sulphidy, putrid	
NCIB 10752 (from chicken)	NH_3, H_2S, turnipy	H_2S, sour, sweaty, fruity	
Pseudomonas perolens	Slightly potato sack odour	Strong potato sack odour	Pyrazine derivative (2 methoxy-3-isopropyl pyrazine)
Pseudomonas fragi			
NCIB 10723 (from milk)	Fruity	Strong, fruity	Esters of Acetic, Butyric, Propionic, Hexanoic Acids
NCMB 1789 (from fish)			
MJT/F5/102 (from chicken)			
NCIB 10476 (from cottage cheese)			
Alteromonas putrefaciens			
NCIB 10471 (from butter)	Slightly sweaty	Sweaty, sour strong sulphidy stale vegetables NH_3, TMA, putrid	NH_3, TMA, H_2S $(CH_3)_2S$; $CH_3{\cdot}SH$ etc.
NCMB 1735 (from stale haddock)			
NCIB 10761 (from poultry)	Slightly faecal		
Moraxella sp.			
No. 78 (from cod fillets)	No change	No change	—

TABLE 8. Organoleptic descriptions[a] of various concentrations of volatile sulphur compounds compared with those from spoiling white fish stored in ice

Compound	Concentration (parts 10^{-12}) in aqueous soln	Odour description	Concentration (parts 10^{-12}) in naturally spoiling white fish after 10 d at 0°C	Organoleptic (odour) changes in white fish stored at 0°C
H_2S	20	No odour		0–6 d: no marked change
	40	Slight H_2S—threshold value	150	
	80	*Medium strong H_2S*		7–10 d: slight musty
	500	Strong H_2S		
$(CH_3)_2S$	0·1	No odour		11–15 d: *sour*, fruity *H_2S sulphidy stale cabbage water, NH_3* and TMA
	0·50	Slight sour trace and sulphide—threshold value	20	
	0·70	Slightly *cabbage water*		
	1·50	Strong sulphidy		
CH_3SH	0·01	No odour		
	0·05	Very slightly sour—threshold value		
	0·10	Slightly musty and sour		
	0·50	Slight *cabbage water*, leeks	120	
	2·00	Sharp strong stale cabbage water		
	100·00	Metallic, cooked meat, sulphidy		

[a] Threshold values are in italics.

TABLE 9. Shelf life of various species of fish stored in ice[a]

Area	Species	Shelf life (d)
Northern temperate waters	Marine	
	Fatty Scotch summer herring	2–4
	Lean Norwegian herring	12
	Mackerel	6
	Whiting	9
	Cod, haddock	12
	Sole, plaice	7–18
	Halibut	21
	Freshwater	
	Perch	15–17
	Trout	9–11
India	Marine	
	Pomfret	7–9
	Horse mackerel	10–45
	Bonga	27
	Perch	30–45
	Freshwater	
	Nile perch, Mrigal carp	29
W. Africa	Marine	
	Sea bream	26
E. Africa	Freshwater	
	Kariba bream	15
	Tilapia sp.	28

[a] After Shewan (1976).

40 d in ice and Nile perch and Mrigal carp (both freshwater species) for up to 29 d, whereas our freshwater perch remain edible for about 17 d and trout 9–11 d. The most obvious explanation is that in such species the psychrophilic bacteria are absent as members of the initial floras of the fish and are only gradually introduced with the ice.

Possible Reasons for Differing Spoilage Rates

It is probable that the defence mechanisms, more powerful in some species of fish than in others, might still operate for some time after death. Certainly histological examination of the integument of various species of fish (Figs 4–6) shows that there are considerable differences in their structures and their physical properties. Plaice, which as already mentioned keeps well, besides having a copious covering of mucus, has a robust well-defined epidermis and dermis, whereas whiting, a quick spoiler, has an extremely poor dermis with a very fragile and easily removable epidermis. Mechanical damage to whiting during handling on

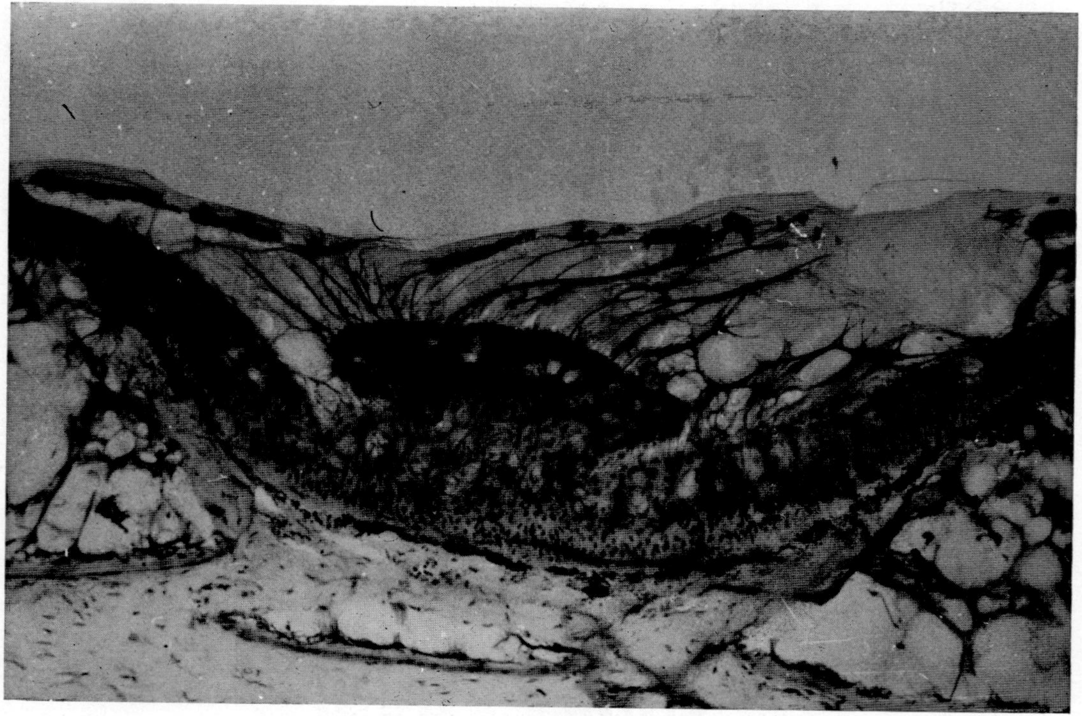

FIG. 4. Fresh plaice showing mucus (Giemsa) × 100.

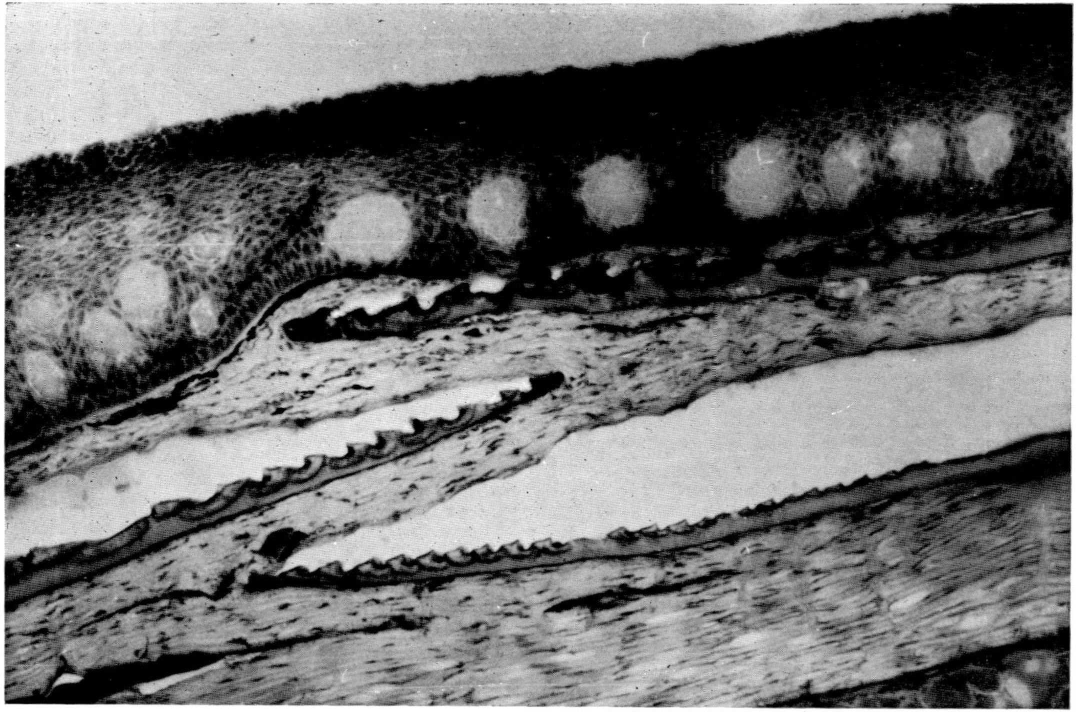

FIG. 5. Fresh cod skin (Alcian blue/basic fuchsin) × 150.

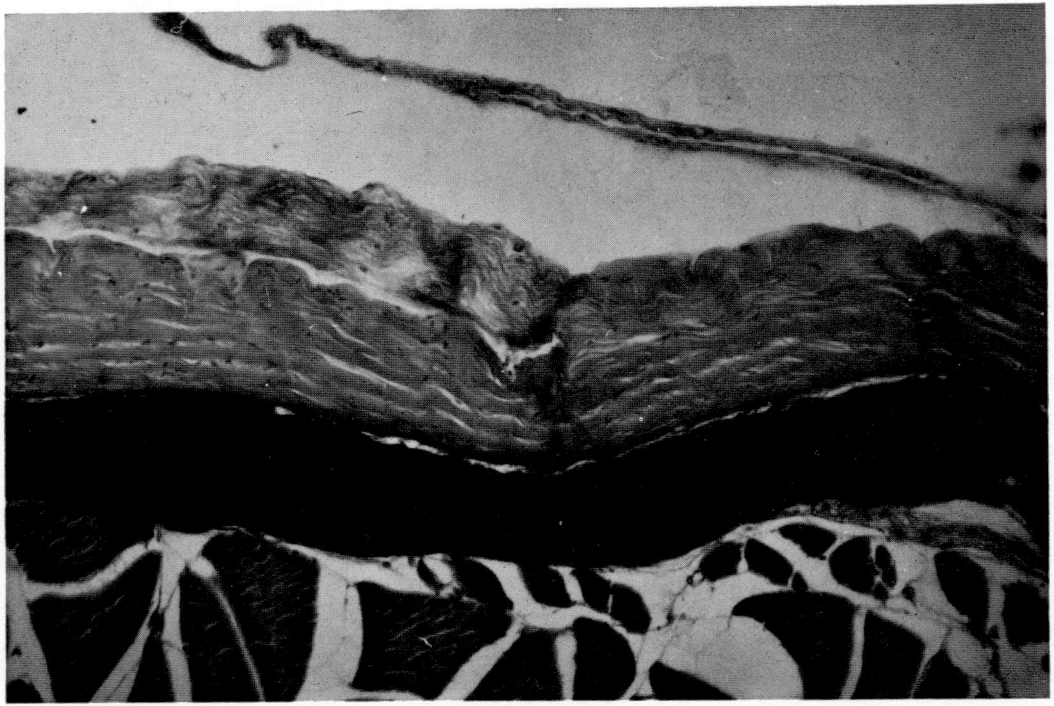

FIG. 6. Fresh blue whiting skin (Giemsa) × 150.

board ship or on shore is therefore very great resulting in rapid spoilage. Blue whiting, a recently exploited species, has similarly a poor integument and also keeps poorly.

There may well be other considerations to explain the differing keeping qualities of the various fish species. Plaice has been shown to possess a powerful lysozyme (Murray and Fletcher, 1976) whereas cod has none. Lysozyme, effective against some Gram positive bacteria, requires the help of chelating agents and surfactants to lyse Gram negative organisms *in vitro*. It is possible that cellular components in plaice slime may provide the necessary cofactors to the enzyme to broaden its activity against microbial invasion. Certain antibacterial properties have already been ascribed to some fish slimes (Zotov, 1958; Liguori *et al.*, 1963).

References

Barnes, E. M. & Thornley, M. J. (1966). The spoilage flora of eviscerated chicken stored at different temperatures. *Journal of Food Technology* **1,** 113–119.

Forster, J. (1887). Über einige Eigenschaften leuchtender Bakterien. *Zentralblatt für Bakteriologie* **2,** 337–340.

Forster, J. (1892). Über die Entwicklung von Bakterien bei niederen Temperaturen. *Zentralblatt für Bakteriologie* **12,** 431–436.

Gillespie, N. C. & Macrae, I. C. (1975). The bacterial flora of some Queensland fish and its ability to cause spoilage. *Journal of Applied Bacteriology* **39,** 91–100.

Harrison-Church, C. (1973). Unpublished results. Torry Research Station, Aberdeen.

Herbert, R. A. & Shewan, J. M. (1975). Precursors of the volatile sulphides in spoiling North Sea cod (*Gadus morhua*). *Journal of the Science of Food and Agriculture* **26,** 1195–1202.

Herbert, R. A., Ellis, J. R. & Shewan, J. M. (1975). Isolation and identification of the volatile sulphides produced during chill-storage of North Sea cod (*Gadus morhua*). *Journal of the Science of Food and Agriculture* **26,** 1187–1194.

Hobbs, G., Cann, D. C., Wilson, B. B. & Horsley, R. W. (1971). The bacteriology of scampi (*Nephrops norvegicus*). III. Effect of processing. *Journal of Food Technology* **6,** 233–251.

Ienistea, C., Chitu, M. & Roman, A. (1970). Les bactéries proteolytiques des viandes réfrigérées de bovins. *Archives Roumaines de Pathologie Experimentale et de Microbiologie* **29,** 305–313.

Lee, J. S. & Harrison, J. M. (1968). Microbial flora of Pacific hake (*Merluccius productus*). *Applied Microbiology* **16,** 1937–1938.

Liguori, V. R., Ruggieri, G. D., Baslow, M. H., Stempien, M. F. & Nigrelli, R. F. (1963). Antibiotic and toxic activity of the mucus of the Pacific striped bass, *Grammistes sexlineatus*. *American Zoologist* **31,** p. 546: Abstract No. 302.

Morita, R. (1975). Psychrophilic bacteria. *Bacteriological Reviews* **39,** 144–167.

Murray, C. K. & Fletcher, T. C. (1976). The immunohistochemical localization of lysozyme in plaice (*Pleuronectes platessa* L.) tissues. *Journal of Fish Biology* **9,** 329–334.

Rogers, R. E. & McCleskey, C. S. (1957). Bacteriological quality of ground beef in retail markets. *Food Technology* **11,** 318–320.

SCHMIDT-NIELSON, S. (1902). Über einige psychrophile Mikroorganismen und ihr Vorkommen. *Zentralblatt für Bakteriologie, Parasitenkunde, Infektionskrankheiten und Hygiene* Abteilung II **19,** 145–147.

SCHULTZE, W. D. & OLSON, J. C. (1960). Studies on psychrophilic bacteria. I. Distribution in stored commercial dairy products. *Journal of Dairy Science* **43,** 346–350.

SHEWAN, J. M. (1974). The biodeterioration of certain proteinaceous foodstuffs at chill temperatures. In *Industrial Aspects of Biochemistry* (Spencer, B., ed.), pp. 475–490. Federation of European Biochemical Societies.

SHEWAN, J. M., HOBBS, G. & HODGKISS, W. (1960). The *Pseudomonas* and *Achromobacter* groups of bacteria in the spoilage of marine white fish. *Journal of Applied Bacteriology* **23,** 463–468.

SHEWAN, J. M. (1976). The bacteriology of fresh and spoiling fish and the biochemical changes induced by bacterial action. In *Proceedings of Tropical Institute Conference on the Handling, Processing and Marketing of Tropical Fish,* pp. 51–66.

SHIBATA, T. (1977). Enzymological studies on the glycolytic system in the muscles of aquatic animals. *Memoirs of the Faculty of Fisheries, Hokkaido University* **24,** 1–80.

SIMIDU, U., KANEKO, E. & AISO, K. (1969). Microflora of fresh and stored flatfish *Kareius bicoloratus. Bulletin of the Japanese Society of Scientific Fisheries* **35,** 77–82.

TOMPKIN, R. B. (1973). Refrigeration temperature as an environmental factor influencing the microbial quality of food—a review. *Food Technology* **27,** 54–58.

ZOTOV, A. F. (1958). Studies on antibacterial characteristics of extracts from fish. I and II. *Voprosy Ikhtiologii* **10,** 120–123; **11,** 188–191.

Psychrotrophs and their Effects on Milk and Dairy Products

B. A. Law, Christina M. Cousins, M. Elisabeth Sharpe and F. L. Davies

National Institute for Research in Dairying, University of Reading, Shinfield, Reading, Berkshire, England

Introduction

Raw milk is bulked and stored at $\leqslant$ 4°C by the producer before collection and often held in insulated silos for a further period before processing. This refrigeration prevents obvious spoilage due for example to lactic acid bacteria, but the microflora of the stored milk becomes dominated by psychrotrophic bacteria (Thomas, 1974; Cousins *et al.*, 1977). In the dairy industry, psychrotrophs are defined as those micro-organisms which can multiply at a temperature of 7°C or less, irrespective of optimal growth temperature. The growth of these psychrotrophs in stored raw milk may affect the efficiency with which the milk can be utilized and also the quality and flavour of milk products.

Methods

Counting methods

Total colony count of milk

From appropriate decimal dilutions made in quarter strength Ringer's solution 1 ml volumes were plated in duplicate using Yeastrel milk agar (Oxoid). Colonies were counted after incubation of plates at 30°C $\pm$ 1°C for 72 $\pm$ 2 h.

Psychrotrophic colony count of milk

Plates were prepared as above with incubation at 5°C $\pm$ 0·5°C for 10 d.

Syringing technique

Using a sterile disposable 5 ml syringe fitted with a 0·8 $\times$ 40 mm hypodermic needle, 5 ml milk was drawn up and expelled three times, each expulsion taking about 1 s (Te Whaiti and Fryer, 1977).

Direct microscopic clump count

Using a metal syringe preset for repeated accurate measurement of 0·01 ml volumes (Applied Research Institute, 2E, 23rd Street, New York 10010, USA), 0·01 ml milk was spread over 1 cm^2 of a glass microscope slide. After drying and staining with Newman's stain, clumps of bacteria and single bacterial cells in representative portions of the film were counted microscopically at a magnification of 900 ×. The clump count ml^{-1} of milk was determined from the average count $field^{-1}$ and the microscope factor. For details of preparation and staining of the film, counting and determination of the microscope factor: see Anon. (1968) and Anon. (1972).

Media for detecting lipolytic and proteolytic activity of psychrotrophs

These media are diagnostic and can be used for the enumeration and isolation of lipolytic and proteolytic bacteria from natural habitats, or for detecting extracellular production of these enzymes by pure cultures.

Lipolysis

Victoria blue butter fat agar. Described by Fryer *et al.* (1967) and Lawrence *et al.* (1967).

Victoria blue base. Boil 5 g of Victoria Blue (BDH, Poole, England) in 500 ml distilled water. Add 10% (w/v) NaOH until brown (about 5 ml). Filter off the precipitate, wash it with water containing a few drops of ammonia, and dry on the filter paper at 30°C.

Fat substrate. Boil fresh unsalted butter for 30 min with an equal volume of distilled water (pH 7·0). Using a warm separating funnel, run off the water phase. Centrifuge the butter oil at a temperature of 70°C for 20 min at 2500 r min^{-1} to separate further. The butter oil should then be clear.

Add Victoria Blue base powder to the melted butter oil while stirring, until it is dark red. Hold at 100°C for 30 min, filter through paper at 30°C. Store in small stoppered tubes at 4°C and use as required.

Preparation of plates. Melt sufficient Victoria Blue butter fat (VBBF) and into this immerse circles of lens tissue 10 S (J. Green, Maidstone, England), cut to just less than the size of a petri dish. Cool plates to 4°C to set, then cover with 15 ml melted nutrient agar, and allow to set.

Rapid method for quantitative estimation of microbial lipases. Use a 1% emulsion of trioctanoin and the slide method described by Lawrence *et al.* (1967). Trioctanoin is used in preference to tributyrin as the latter triglyceride may also be hydrolysed by esterases. This method is used for pure cultures and cell-free extracts.

Proteolysis

Skim milk agar. Used to detect caseinolytic activity and described by Harrigan and McCance (1976). To 50 ml melted 2% plain agar at 50°C add 25 ml warm sterilized skim milk (or skim milk powder resuspended in water and sterilized). Mix well, pour plates.

Samples. Suitable dilutions of samples (0·1 ml) are spread on the surface of previously dried plates of VBBFA or skim milk agar. Broth cultures are plated directly. Incubate both media for up to 1 week at 25°C. Lipolysis is indicated by the Victoria Blue giving a blue zone under the bacterial growth, and caseinolysis by a clearing zone.

Identification methods. GNR isolates were identified according to the scheme of Hendrie and Shewan (1966), Thornley (1967), Shewan *et al.* (1960) and Stanier *et al.* (1966).

Enumeration and Growth of Psychrotrophic Bacteria in Raw Milk

For raw milk in the UK there are no statutory standards for either total or psychrotrophic bacterial counts. However, the pour plate method using Yeastrel milk agar (YMA) is specified for both counts (Anon., 1968) with incubation of plates at 5–7°C for 7–10 d for the psychrotroph count (PC). In the USA, Plate Count Agar (PCA) is used and the time–temperature combination for PC is 7°C $\pm$ 1°C for 10 d (Anon., 1972). The long incubation period means that the standard PC is used more for research, survey and investigational purposes than for routine control in industry.

Most psychrotrophs of concern in milk and milk products can form colonies on YMA and PCA incubated at 30–32°C and, therefore, they contribute to the total colony count (TC). Storage of milk at $\leqslant$ 7°C is selective for psychrotrophs and the increase in TC observed during refrigerated storage of raw milk is caused by multiplication of the psychrotrophic microflora. Examples of the changes in TC and PC occurring in individual farm milks stored at 5°C are shown in Fig 1 a, b; after 3–4 d the PC approached or equalled the TC. The difference between the two milks, having similar initial TC, in the rate of increase in bacterial counts illustrates the difficulty of predicting from the initial count the bacteriological quality after refrigerated storage.

Milk samples from farm collection tankers, bulks of six or seven farms' output, were stored at 5 and 7°C. The more rapid increase in the

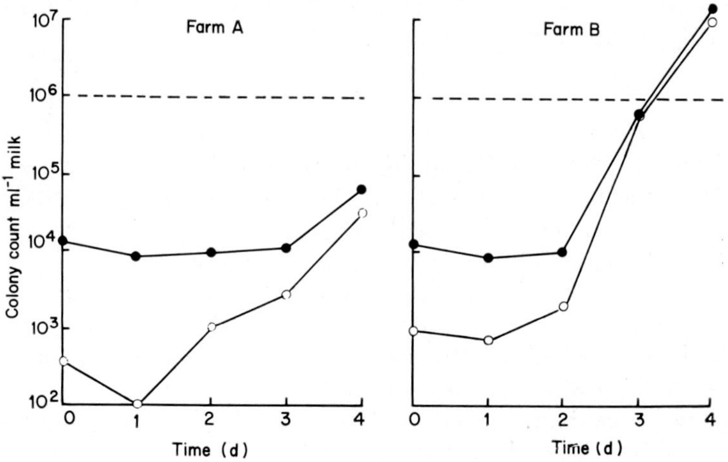

FIG. 1. Total (●) and psychrotroph (○) colony counts of individual farm bulk tank milks during storage at 5°C.

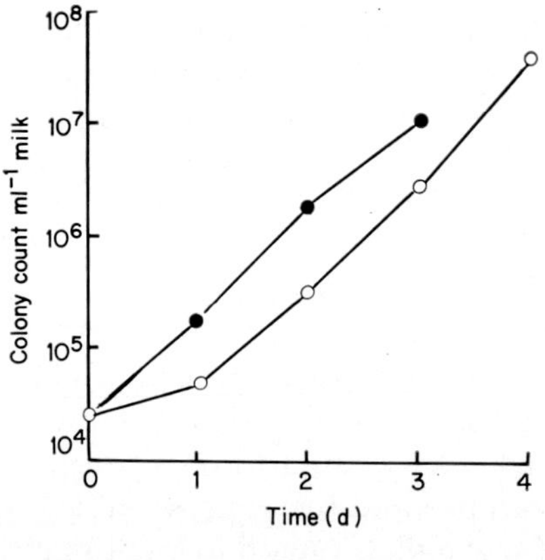

FIG. 2. Psychrotroph colony counts of farm collection tanker milk during storage at 5°C (○) and 7°C (●); means of eight samples.

mean PC at 7°C as compared with that at 5°C (Fig. 2) reduces by about 1 d the time taken to reach a given count.

Effect of cell clumping on colony count

It is well recognized that the plate count underestimates the number of viable bacterial cells in raw milk, one reason being the presence of chains, groups or clumps of cells. Te Whaiti and Fryer (1977) have reported that syringing diluted raw milk increased the colony count by as much as ten-fold. Their results indicated that the increase was particularly marked where Gram negative rods were dominant in the microflora, and that large clumps were formed when *Pseudomonas* spp. were grown at low temperature with agitation of the culture during growth. It is not clear to what extent clump formation in raw milk is dependent on types of micro-organisms present, the agglutinating properties of the milk, the time and the temperature of storage, or degree of agitation (or aeration) of the milk. Furthermore, no information was available on the extent of disruption of clumps by syringing stored raw milk.

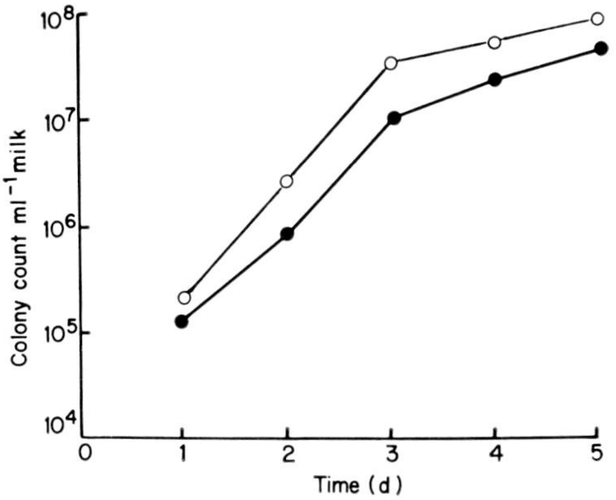

FIG. 3. Total counts of silo tank milks during storage at 7°C before syringing (●) and after syringing (○); means of eight samples.

Samples of milk from silo tanks containing 80–120 × 10^3 l were taken after the milk had been stored for 24 h at 4–5°C. The samples were then stored at 7°C ± 0·5°C in the laboratory, plated initially and at daily intervals for TC before and after syringing in a standard manner. The mean results for samples from eight tanks are shown in Fig. 3. The TC of unsyringed individual samples varied initially (39 × 10^3–44 × 10^4 ml^{-1}) and in the rate of increase during the storage period, in spite of the large

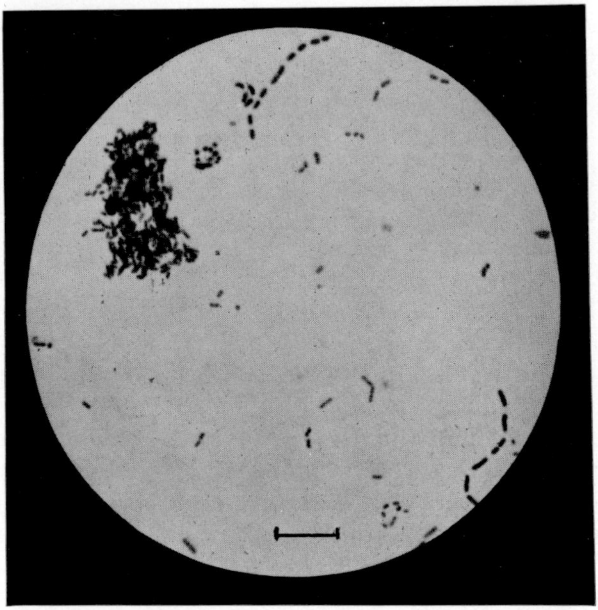

FIG. 4. Light micrograph of a portion of a stained smear of milk stored for 5 d at 5°C; (a) before syringing, (b) after syringing the milk. Bar marker represents 10 μm.

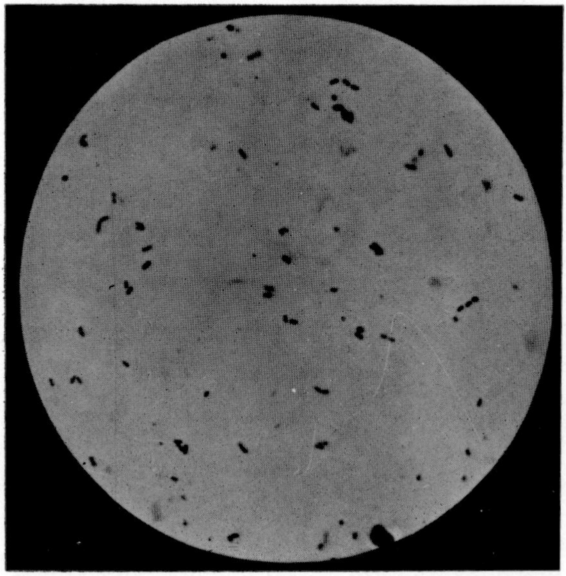

FIG. 4 (b)

bulks of milk from which the samples were drawn. The effects of syringing were not consistent, the resulting increases in TC ranging from $< 2\times$ to $> 10\times$. On average, syringing had least effect initially, thereafter the increase was about four-fold.

Direct microscopic clump counts of the milks before and after syringing showed that although most larger clumps and chains were disrupted by syringing, pairs and a few short chains remained (Fig. 4), occasionally large clumps were observed. Thus, although syringing milk before plating undoubtedly increases the colony count so that it more nearly approaches the total number of viable bacterial cells, these will still be underestimated in stored raw milk. It has not yet been determined whether PC on syringed milks correlate better than PC on unsyringed milk with enzyme production in the milk and subsequent enzymic activity in heated milk or products made from it.

By enumeration of psychrotrophs in raw milk and determination of growth rates, requirements and conditions necessary for limiting numbers to levels unlikely to cause problems in processed milk and products can be determined.

Lipolytic and Proteolytic Psychrotrophs in Raw Milk

Most psychrotrophic bacteria isolated from raw milk are Gram negative rods (GNR) and many of these produce lipases (Law *et al.*, 1976) and

TABLE 1. Lipolytic and proteolytic psychrotrophic bacteria isolated from raw milk

Organism		Lipolytic	Proteolytic
Pseudomonas	*fluorescens*[a]	+ and −	+
	putida[a]	+	+
	aureofaciens	+	+
	fragi[a]	+	+
Acinetobacter sp.[a]		+	+
Aeromonas sp.[a]		+	+
Xanthomonas sp.		±	+
Cytophaga sp.		NT	+
Coliforms		+	+
Proteus sp.		+	+

±, weakly active.
NT, not tested.
[a] most commonly occurring.

proteinases (Law *et al.*, 1977). The highest incidence of lipolytic and proteolytic activity is found among the pseudomonads (Table 1) although some strains of *Ps. fluorescens* are proteolytic but not lipolytic when tested on Victoria Blue Butterfat Agar. The psychrotrophs themselves are killed by heat treatment of the raw milk but their lipases and proteinases survive and may cause spoilage of products manufactured from the heat-treated milk.

Heat resistant lipases and their effect on Cheddar cheese flavour

The production of heat resistant lipases by GNR is now well documented (e.g. Stadhouders and Mulder, 1958; Kishonti and Sjöstrom, 1970; Law *et al.*, 1976) and the inherent heat stability of these enzymes appears to be enhanced by a protective effect in milk (Fig. 5). Because the lipolytic activities of individual strains of psychrotrophic GNR are extremely variable (Table 2) it is difficult, from a knowledge of the num-

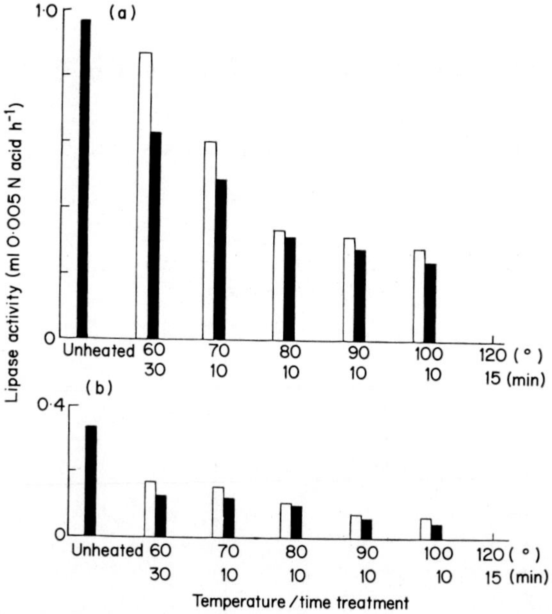

FIG. 5. Effect of heat treatment on the extracellular lipase of *Pseudomonas fluorescens* AR11 (a) and *Ps. fragi* ER 25 (b) in 0·1 M phosphate buffer, pH 6·6 (■), and in separated milk at the same pH (□). The lipases were assayed at 30°C for 2 h with emulsified butterfat as substrate by the method of Alford and Pierce (1963). (Data from Law *et al.*, 1976.)

TABLE 2. Lipolytic activities of bacteria isolated from 3 samples of raw milk (A, C and E)

Bacteria	No. of strains giving lipolytic zone [a] with diam. < 1·0 mm			1·0–2·9 mm			3·0–5·0 mm		
	A	C	E	A	C	E	A	C	E
Pseudomonas fluorescens	–	6	–	7	10	1	1	2	–
Ps. putida	7	–	1	6	2	1	–	–	–
Ps. fragi	1	–	–	4	3	8	–	–	1
Acinetobacter spp.	–	–	1	3	–	2	–	–	–
Aeromonas hydrophila	0	1	3	1	5	9	–	1	–

[a] Determined by the method of Lawrence *et al.* (1967).
–, none isolated
Data from Law *et al.* (1976).

bers and species present, to predict the potential of the milk to yield products liable to rapid spoilage. For this purpose a simple rapid assay for lipase in milk is required but a sufficiently sensitive test has not yet been developed. However, some indication of the numbers of psychrotrophs required to produce off-flavours in cheese has been gained from experiments at this Institute (Law *et al.*, 1976) in which cheeses were made with defined bacterial floras (Mabbitt *et al.*, 1959; Chapman *et al.*, 1966). Lipolytic rancidity developed in maturing cheeses made from pasteurized (71°C 17 s^{-1}) milks derived from stored raw milk previously inoculated with mixed GNR reference floras (Fryer *et al.*, 1966) or a single strain of *Ps. fluorescens* (AR11; NCDO 2085). Rancidity developed within 2–4 months if the raw milk counts had reached *ca.* 10^7 cfu ml^{-1} (Fig. 6), and the off-flavour was characterized by a soapy taste coinciding with the production in the cheeses of free fatty acids (butyric and higher acids) at concentrations from 3–10 times higher than those in control cheeses made from stored low count uninoculated milk. Additional trials have confirmed that psychrotrophic counts in the region of 10^6 to 10^7 cfu ml^{-1} in raw market milk lead to rancidity development in cheeses even though the cheesemilks were pasteurized before manufacture. If the heat resistance of the two pseudomonad lipases shown in Fig. 5 are typical of psychrotroph lipases generally, then heat treatment of milk beyond the temperatures compatible with its use in the manufacture of most cheese varieties would be required to inactivate these enzymes.

Heat resistant proteinases and their effect on UHT-sterilized milk

Most of the psychrotrophs reported to produce heat resistant proteinases have been identified as species of *Pseudomonas* (e.g. Adams *et al.*, 1975;

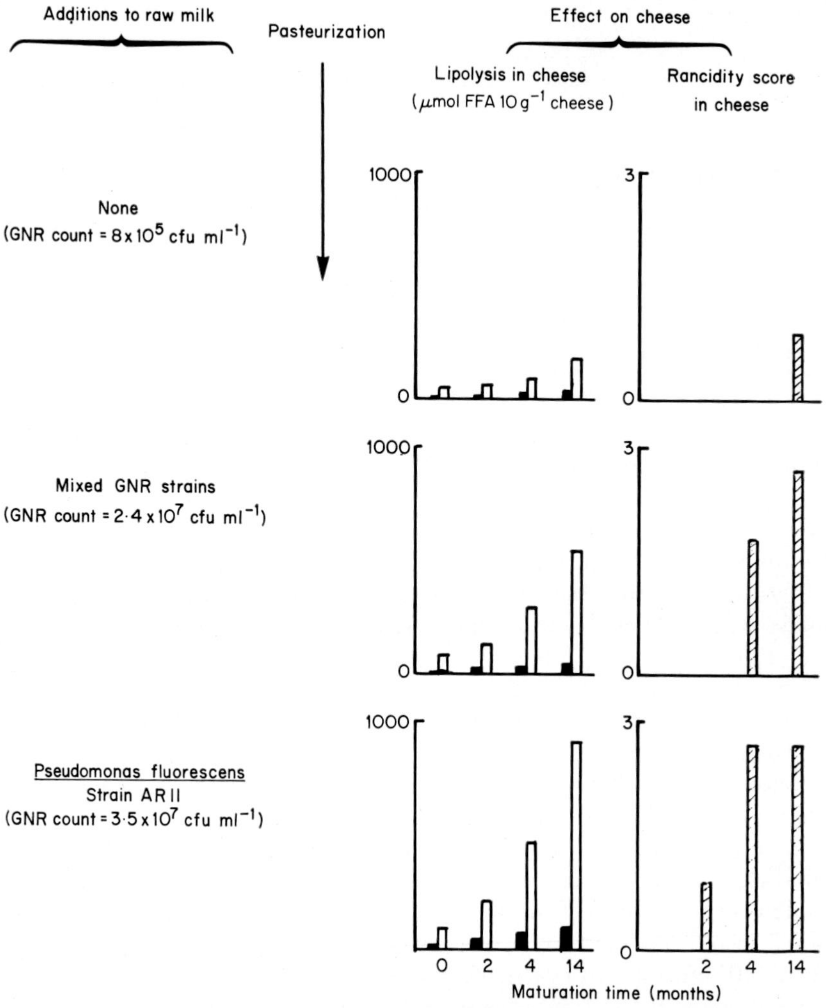

FIG. 6. Effect of the growth of lipolytic psychrotrophic Gram negative rods (GNR) in stored milk on free fatty acid (FFA) concentrations and flavour in Cheddar cheese. Single herd raw milks from this Institute were stored at 10°C for 48 h with or without an inoculum of approx. 1×10^4 cfu ml^{-1} of lipolytic GNR (mixed strains, or *Pseudomonas fluorescens*, strain AR11). The milks were pasteurized to kill the GNR before being used for cheesemaking. Butyric acid (■); higher fatty acids (□). (Data from Law *et al.*, 1976.)

Law *et al.*, 1977) but other genera also contain caseinolytic species (e.g. *Flavobacterium*, *Acinetobacter*, *Aeromonas*, *Xanthomonas*, *Proteus*, coliforms). The proteinase produced by *Ps. fluorescens* strain AR11 is resistant to treatment at time/temperature combinations used for UHT-sterilization of milk (Fig. 7). Stored UHT milk gelled after 12–56 d

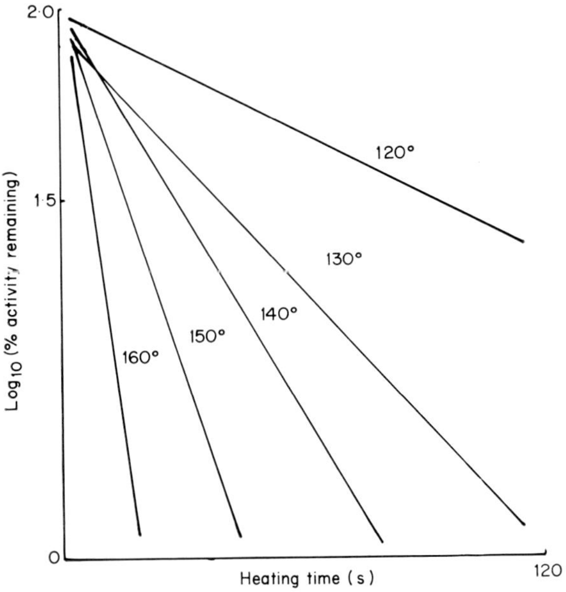

FIG. 7. Heat inactivation of the proteinase of *Pseudomonas fluorescens*, AR11 in 0·05 M phosphate buffer, pH 6·5 at various temperatures. (Data from Alichanidis and Andrews, 1977.)

when it was prepared from raw milks inoculated with AR11 and held for 2–3 d at 7·5°C before processing. The gelled UHT milk samples had contained 10^6–10^7 cfu AR11 ml^{-1} and although these were killed by the heat treatment, the proteinase continued to act during storage of the UHT milk at room temperature; β- and κ-casein fractions were extensively degraded (Law *et al.*, 1977) although breakdown of as, casein was less marked (Table 3).

Other effects of psychrotroph proteinases

The action of proteinases produced by psychrotrophs during the cold storage of raw milk can lead to changes which affect the behaviour of the milk during cheese manufacture. Cousin and Marth (1977) showed that

TABLE 3. Breakdown of the major casein fractions in stored UHT-sterilized milk by heat resistant proteinases produced in raw milk by *Pseudomonas fluorescens* AR11

No. *Ps. fluorescens* in stored milk before sterilization	Time taken for gelation after sterilization (d)	% casein breakdown at gelation time		
		αS_1	β	κ
8×10^5	56[a]	0	0	T
8×10^6	56	0	30	50
5×10^7	12	22	78	100

[a] No gelation occurred during experimental period, stored at 20°C.
T, trace.
Data from Law *et al.* (1977).

rennet coagulation times were shorter during Cheddar cheesemaking with milks which had been inoculated with psychrotrophic strains of *Pseudomonas* and *Flavobacterium* spp. before heat treatment. Chapman *et al.* (1976) also reported changes in coagulation times and in the rigidity of rennet gels in cheesemilks derived from stored raw milk that had contained large numbers of caseinolytic psychrotrophic bacteria. McCaskey and Babel (1966) showed that under laboratory conditions, losses of N into whey during milk coagulation by rennet were increased in milks containing high psychrotrophic counts. This would be expected to result in reduced cheese yields of commercial significance and Feuillat *et al.* (1976) have demonstrated such losses in the form of polypeptide breakdown products from casein during the manufacture of soft cheese.

Suppression of psychrotrophs in raw milk

Since the enzymes of psychrotrophs cannot be destroyed by heat treatment the only means of preventing their production is to reduce or stop the growth of the psychrotrophs themselves. This can be done by observing stringent precautions to keep the milk at or below 4°C during storage; killing the psychrotrophs by heating (63°C for 15 s) the milk before storage (Schipper, 1975); suppressing or destroying the psychrotrophs by making use of the naturally occurring lactoperoxidase system in milk (e.g. Björck *et al.*, 1975).

Psychrotrophic Sporeforming Bacteria

In addition to Gram negative psychrotrophs, raw milk contains psychrotrophic sporeforming rods (SFR), commonly of the genus *Bacillus* (Shehata and Collins, 1971) and more rarely of the genus *Clostridium* (Bhadsavle *et al.*, 1972).

Isolation

Psychrotrophic SFR are generally slower growing than their Gram negative counterparts (Langeveld *et al.*, 1973) but can be readily isolated from milk if the GNR are first inactivated by a heat treatment (e.g. 75°C for 10 min) which the spores can survive. Heat-treated milk incubated at 5°C for 14 d and plated on a rich medium such as Tryptic Soy Agar (BBL) will generally yield colonies of SFR within 7 d at 5°C. The number and variety of such bacteria obtained can be increased by enriching the heat treated milk with Tryptic Soy Broth (BBL) at a rate of 1 vol milk: 10 vol broth before incubation at 5°C (Parry and Davies, unpublished).

On primary isolation, psychrotrophic SFR frequently appear Gram negative or Gram variable and may remain so through several sub-cultures so that care must be taken to avoid confusion with other genera.

Classification

It is frequently difficult to match psychrotrophic SFR from milk with published descriptions of either mesophilic (e.g. Gordon *et al.*, 1973) or proposed psychrophilic (Larkin and Stokes, 1967) species of *Bacillus*.

Of 76 such isolates obtained from 72 farm bulk milk tanks, 54 most closely resembled *B. circulans* (though sporangia were not always definitely swollen), 7 corresponded to *B. sphaericus* and the remainder represented other species of *Bacillus* as single isolates or could not be classified (Parry and Davies, unpublished). Clearly the temperature limits for growth of these isolates were lower than those published for *B. circulans* or *B. sphaericus*. While most *Bacillus* species have been reported to occur as psychrotrophs in milk (e.g. Shehata and Collins, 1971), isolates resembling the *B. circulans* variants found in this study are probably the most common (e.g. Langeveld *et al.*, 1973; Grosskopf and Harper, 1974).

While it is probable that these organisms represent mesophilic species which have adapted to growth at lower temperature, it is interesting that the species distribution of the psychrotrophs is quite different from that of the mesophilic SFR in milk where *B. subtilis* is generally the most common (Underwood *et al.*, 1974).

Growth and sporulation

Of our isolates, those described as *B. sphaericus* showed the greatest adaptation to low temperature, growing and sporulating on laboratory media between 0°C and 25°C or 30°C whereas the *B. circulans* variants grew between 5°C and 37°C, many failing to sporulate at the lower temperature. In milk at 5°C the most rapidly growing of our psychrotrophic SFR were strains of *B. cereus* with doubling times of about 14 h. Langeveld *et al.* (1973) reported doubling times of about 12 h for psychrotrophic *B. circulans* in milk at 5°C compared to only 5–6 h for pseudomonads.

Heat resistance

Shehata and Collins (1972) suggested that spores of psychrotrophic SFR are less thermoresistant than those of mesophiles. Data obtained by Laine (1970), however, do not support this contention and suggest that sporulation temperature is the more important determinant of heat resistance. Our results with isolates of *B. sphaericus* confirm that spores produced at low temperatures (e.g. 5°C) possess lower thermo-resistance than those produced at higher temperatures (e.g. 25°C) (Davies, 1975).

Commercial significance

Organisms with the ability to survive processing temperatures and grow at refrigeration temperatures clearly possess high spoilage potential. In practice, however, there have been few reports of spoilage due to such bacteria. Overcast and Atmaran (1974) reported that 28% of commercial pasteurized milk samples from Tennessee showed sweet curdling from *B. cereus* within 10 days of refrigerated storage and we found that this organism developing in milk residues of inadequately cleaned road tankers could multiply and cause spoilage of pasteurized milk on refrigeration (Davies, 1975). In general, however, the slow rate of development of psychrotrophic SFR in milk reduces their spoilage potential and in common with GNR, they can usually be suppressed by strict control of refrigeration temperatures.

Acknowledgements

The technical assistance of Miss U. Rodrigues is gratefully acknowledged. We also thank Mr. B. A. Phillips, Mr. H. M. Underwood and Mr. O. T. Parry for their cooperation.

References

Adams, D. M., Barach, J. T. & Speck, M. L. (1975). Heat resistant proteases produced in milk by psychrotrophic bacteria of dairy origin. *Journal of Dairy Science* **58,** 828–834.

Alford, J. A. & Pierce, D. N. (1963). Production of lipase by *Pseudomonas fragi* in a synthetic medium. *Journal of Bacteriology* **86,** 24–29.

ALICHANIDIS, E. & ANDREWS, A. T. (1977). Some properties of the extracellular protease produced by the psychrotrophic bacterium *Pseudomonas fluorescens* strain AR11. *Biochimica et Biophysica Acta* **485,** 424–433.

ANON. (1968). *Methods of Microbiological Examination for Dairy Purposes* BS.4285, London: British Standards Institution.

ANON. (1972). *Standard Methods for the Examination of Dairy Products* 13th edn., Washington D.C.: American Public Health Association.

BHADSAVLE, C. H., SHEHATA, T. E. & COLLINS, E. B. (1972). Isolation and identification of psychrophilic species of *Clostridium* from milk. *Applied Microbiology* **24,** 699–702.

BJÖRCK, L., ROSÉN, C-G., MARSHALL V. & REITER, B. (1975). Antibacterial activity of the lactoperoxidase system in milk against pseudomonads and other Gram negative bacteria. *Applied Microbiology* **30,** 199–204.

CHAPMAN, H. R., MABBITT, L. A. & SHARPE, M. E. (1966). Apparatus and techniques for making cheese under controlled bacteriological conditions. *17th International Dairy Congress* **D,** 55–60.

CHAPMAN, H. R., SHARPE, M. E. & LAW, B. A. (1976). Some effects of low-temperature storage of milk on cheese production and Cheddar cheese flavour. *Dairy Industries International* **41,** 42–45.

COUSIN, M. A. & MARTH, E. H. (1977). Cheddar cheese made from milk that was precultured with psychrotrophic bacteria. *Journal of Dairy Science* **60,** 1048–1056.

COUSINS, C. M., SHARPE, M. E. & LAW, B. A. (1977). The bacteriological quality of milk for Cheddar cheesemaking. *Dairy Industries International* **42** (7), 12–17.

DAVIES, F. L. (1975). Heat resistance of *Bacillus* species. *Journal of the Society of Dairy Technology* **28,** 69–78.

FEUILLAT, M., LE GUENNEC, S. & OLSSON, A. (1976). Proteolysis of refrigerated milk and effects on soft cheese yield. *Lait* **56,** 521–536.

FRYER, T. F., SHARPE, M. E. & REITER, B. (1966). The microflora of Cheddar cheese made aseptically from milk containing a standard raw milk reference flora. *17th International Dairy Congress* **D,** 61–66.

FRYER, T. F., REITER, B. & LAWRENCE, R. C. (1967). Methods for isolation and enumeration of lipolytic organisms. *Journal of Dairy Science* **50,** 477–484.

GORDON, R. E., HAYNES, W. C. & PANG, C. H-W. (1973). The Genus *Bacillus. U.S. Department of Agriculture* Handbook No. 427, Washington D.C.

GROSSKOPF, J. C. & HARPER, W. J. (1974). Isolation and identification of psychrotrophic sporeformers in milk. *Milchwissenschaft* **29,** 467–470.

HARRIGAN, W. F. & MCCANCE, M. E. (1976). In *Laboratory methods in Food and Dairy Microbiology*, London and New York: Academic Press.

HENDRIE, M. S. & SHEWAN, J. M. (1966). The identification of certain *Pseudomonas* species. In *Identification Methods for Microbiologists* (Gibbs, B. M. & Skinner, F. A., eds). Society for Applied Bacteriology, Technical Series No. 1, Part A. London and New York: Academic Press, pp. 1–8.

KISHONTI, E. & SJÖSTRÖM, G. (1970). Influence of heat resistant lipases and proteases in psychrotrophic bacteria on product quality. *XVIII International Dairy Congress* **1E,** B.3.

LAINE, J. J. (1970). Studies on psychrophilic bacilli of food origin. *Annales Academiae Scientiarum Fennicae.* A. IV. *Biologica* **169,** 1–36.

LANGEVELD, L. P. M., CUPERUS, F. & STADHOUDERS, J. (1973). Bacteriological aspects of the keeping quality at 5° of reinfected and non-reinfected pasteurized milk. *Netherlands Milk and Dairy Journal* **27,** 54–65.

LARKIN, J. M. & STOKES, J. L. (1967). Taxonomy of psychrophilic strains of *Bacillus. Journal of Bacteriology* **94,** 889–895.

LAW, B. A., ANDREWS, A. T. & SHARPE, M. E. (1977). Gelation of ultra-high-temperature-sterilized milk by proteases from a strain of *Pseudomonas fluorescens* isolated from raw milk. *Journal of Dairy Research* **44,** 145–148.

LAW, B. A., SHARPE, M. E. & CHAPMAN, H. R. (1976). The effect of lipolytic Gram negative psychrotrophs in stored milk on the development of rancidity in Cheddar cheese. *Journal of Dairy Research* **43,** 459–468.

LAWRENCE, R. C., FRYER, T. F. & REITER, B. (1967). Rapid method for the quantitative estimation of microbial lipases. *Nature, London* **213,** 1264–1265.

MCCASKEY, T. A. & BABEL, F. J. (1966). Protein losses in whey as related to bacterial growth and age of milk. *Journal of Dairy Science* **49,** 697.

MABBITT, L. A., CHAPMAN, H. R. & SHARPE, M. E. (1959). Making Cheddar cheese on a small scale under controlled bacteriological conditions. *Journal of Dairy Research* **26,** 105–112.

OVERCAST, W. W. & ATMARAN, K. (1974). The rôle of *Bacillus cereus* in sweet curdling of fluid milk. *Journal of Milk and Food Technology* **37,** 233–236.

SCHIPPER, C. J. (1975). IDF *Annual Bulletin, Document No. 86,* Brussels: IDF.

SHEHATA, T. E. & COLLINS, E. B. (1971). Isolation and identification of psychrophilic species of *Bacillus* from milk. *Applied Microbiology* **21,** 466–469.

SHEHATA, T. E. & COLLINS, E. B. (1972). Sporulation and heat resistance of psychrophilic strains of *Bacillus. Journal of Dairy Science* **55,** 1405–1409.

SHEWAN, J. M., HOBBS, G. & HODGKISS, W. (1960). A determinative scheme for the identification of certain Genera of Gram negative bacteria, with special reference to the Pseudomonadaceae. *Journal of Applied Bacteriology* **23,** 379–390.

STADHOUDERS, J. & MULDER, H. (1958). Micro-organisms involved in the hydrolysis of fat in the interior of cheese. *Netherlands Milk and Dairy Journal* **12,** 238–264.

STANIER, R. Y., PALLERONI, N. J. & DOUDOROFF, M. (1966). Aerobic pseudomonads: a taxonomic study. *Journal of General Microbiology* **43,** 159–271.

TE WHAITI, I. E. & FRYER, T. F. (1977). The enumeration of bacteria in refrigerated milk. *New Zealand Journal of Dairy Science and Technology* **12,** 51–57.

THOMAS, S. B. (1974). The influence of the refrigerated farm bulk milk tank on the quality of the milk at the processing dairy. *Journal of the Society of Dairy Technology* **27,** 180–187.

THORNLEY, M. J. (1967). A taxonomic study of *Acinetobacter* and related genera. *Journal of General Microbiology* **49,** 211–257.

UNDERWOOD, H. M., MCKINNON, C. H., DAVIES, F. L. & COUSINS, C. M. (1974). Sources of *Bacillus* spores in raw milk. *19th International Dairy Congress* **1E**, 373–374.

Bactericidal Activity of the Lactoperoxidase System against Psychrotrophic *Pseudomonas* spp. in Raw Milk

B. Reiter and Valerie M. Marshall

National Institute for Research in Dairying, University of Reading, Shinfield, Reading, Berkshire, England

Introduction

As so often happens, technical advances tend to create new problems. When milk was collected daily in churns from the farm and the creameries processed the whole day's intake, psychrotrophs were not important spoilage organisms. With the advent of refrigerated bulk tanks on farms, milk is now collected in insulated (but not refrigerated) tankers and stored in insulated silos. Sometimes, as in the off-peak season, the milk is only collected every other day and the storage time in the silo is extended to several days, particularly during holidays to suit labour conditions. It is, therefore, not surprising that psychrotrophic organisms have become important spoilage organisms both in liquid milk and dairy products. Although the bacteria are easily destroyed by heat treatment, their extracellular enzymes—proteases and lipases—are heat resistant and, even if they do not affect the milk quality at the time, they will produce off-flavours and other types of spoilage in dairy products during storage and maturation, e.g. in sterilized milk, butter and cheese (Law *et al.*, 1976, 1977).

It is, therefore, of practical importance to prevent undue multiplication of the psychrotrophs or even reduce their numbers. At refrigeration temperatures, the psychrotrophs tend to multiply after two days and it will be shown in this paper that activation of the lactoperoxidase system drastically reduces their number and can hold them at the same low level for an extended storage time.

The Lactoperoxidase System

Of the naturally occurring inhibitory systems in raw milk (immunoglobulins, complement, lysozyme, lactoferrin, etc.), only the lactoperoxidase (LP) system can be simply activated and reduce the bacterial content of milk (for recent reviews see Reiter, 1976, 1978).

The complete LP system consisting of LP, thiocyanate (SCN^-) and H_2O_2 was first demonstrated to inhibit the multiplication and acid production of some lactic acid streptococci (see review by Reiter and Oram, 1967). Of the three components, LP is always present in bovine milk at sufficient concentrations (up to 30 mg ml^{-1}) but the SCN^- content depends on the feeding regime. It is derived from the metabolism of S-containing amino acids, by detoxification of CN^- (e.g. from clover) and by enzyme hydrolysis of glucosides (e.g. from Brassicaceae and *Raphanii*). The level of SCN^- can be as low as 0·01 mM or as high as 0·25 mM when the cows are on natural pasture (Boulangé, 1959).

The limiting factor is H_2O_2, but lactic acid bacteria produce sufficient H_2O_2 under aerobic conditions to activate the LP system. These organisms are therefore self-inhibitory in milk even in the presence of catalase because LP has a greater affinity than catalase for H_2O_2. The end products of SCN^- oxidation are SO_4^{2-}, NH_4^{1+}, and CO_2; these are non-inhibitory, but an intermediate oxidation product of SCN^- (HOSCN) is now regarded as the active compound (Oram and Reiter, 1966; Hogg and Jago, 1970; Hoogendorn *et al.*, 1977; Aune and Thomas, 1977).

Catalase positive organisms can only be inhibited by the LP system if an exogenous source of H_2O_2 is supplied, either by addition of H_2O_2 itself or better, by enzymic generation (e.g. glucose and glucose oxidase). Catalase-positive, Gram-negative organisms (e.g. coliforms, pseudomonads, salmonellae and shigellae) are not only inhibited, but are killed by the LP system (Björck *et al.*, 1975; Reiter *et al.*, 1976).

The Bactericidal Activity of the LP System against *Pseudomonas* spp. in a Synthetic Medium

The sequence of the concomitant oxidation of SCN^- by the LP/H_2O_2 complex and the killing of bacteria, can be illustrated in a synthetic ammonium salts medium (Reiter *et al.*, 1976) because the complex composition of milk makes the assays required more difficult and time consuming.

The SCN^- is rapidly oxidized and at 2 h the bacterial count is reduced by more than 99% (Fig. 1); although the glucose oxidase generates H_2O_2, no free H_2O_2 can be detected until most of the SCN^- is oxidized.

After 3 h, H_2O_2 accumulates and eventually reaches bactericidal levels after 3 to 4 h. These results indicate that the bacterial catalase does not effectively compete with LP for the H_2O_2 generated by the glucose and glucose oxidase.

In the absence of LP and an oxidizable substrate the bacterial catalase is effective, as shown in Fig. 2. Organisms suspended in a synthetic medium hydrolyse the H_2O_2 generated from 0·1 u ml^{-1} glucose oxidase and the organisms remain viable. Previous work with *E. coli* has shown that when the glucose oxidase is increased to 0·5 u ml^{-1}, the same number of organisms are unable to hydrolyse the H_2O_2 generated at the higher rate and the H_2O_2 rises to bactericidal levels (Reiter *et al.*, 1977).

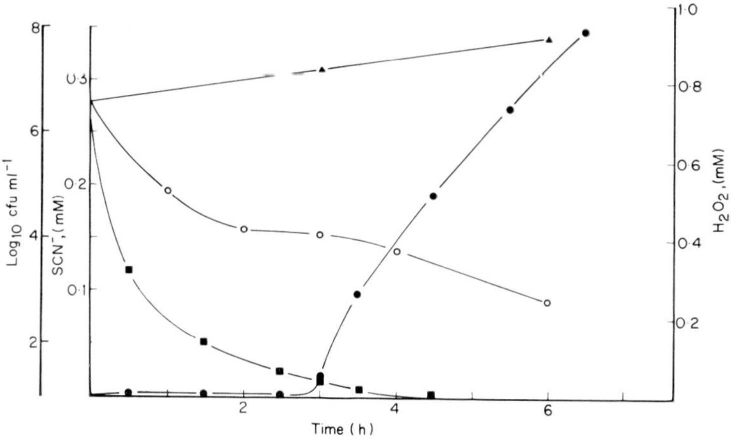

FIG. 1. Relationship between H_2O_2 production, SCN^- oxidation and viable count of *Ps. fluorescens* in a glucose-salts medium (0·3% glucose, pH 6·7) at 30°C. Viable count of *Ps. fluorescens* with no additions (▲); viable count of *Ps. fluorescens* in the presence of SCN^- (0·26 mM), glucose oxidase (0·05 u ml^{-1}) and LP (1·5 u ml) (○); H_2O_2 concentration (●); SCN^- concentration (■).

Since milk contains LP as well as catalase, it was important to determine the level at which catalase could interfere with the bacterial activity of the LP system. Only very high concentrations of catalase were capable of reversing the bactericidal effect (Fig. 3). Such levels of catalase are quite unphysiological, even in highly infected udders, where up to 5 u ml^{-1} of catalase may be present as a consequence of large numbers of leucocytes, the bacterial activity of the LP system would not be reversed.

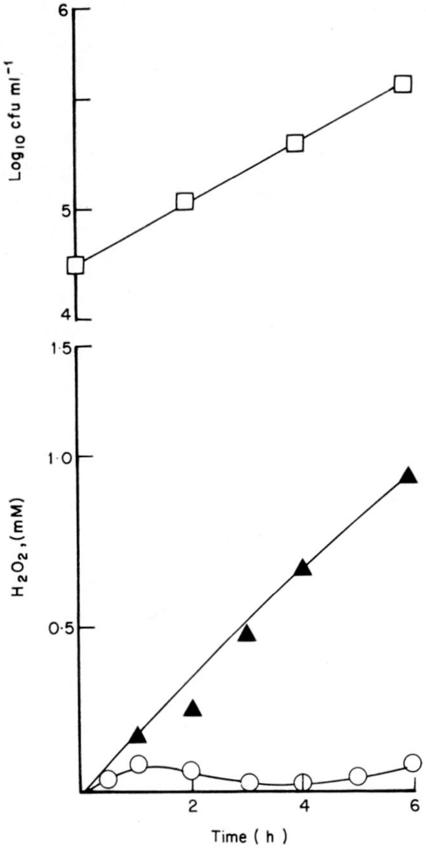

FIG. 2. H_2O_2 generation from glucose oxidase (0·1 u ml^{-1}) in a glucose-salts medium (0·3% glucose) in the presence/absence of bacteria. Bacterial count (■); H_2O_2 concentration in absence of bacteria (▲); H_2O_2 concentration in presence of bacteria (○).

The Nature of the Bactericidal Activity of the LP System

It has been previously shown that in the presence of the LP system *E. coli* is killed after 1½–2 h at which point lysis occurs (see review by Reiter, 1976). *Ps. fluorescens* is affected in the same way (Fig. 4). Up to 30 min, the organisms are capable of recovering when they are removed to a fresh medium after being exposed to the LP system. It has now been shown (Marshall and Reiter, 1976; Marshall, 1978) that the LP system damages the inner membrane within minutes; amino acids and $^{42}K^+$ leak into the medium (Fig. 5) and energy linked transport of glucose (Fig. 6) and

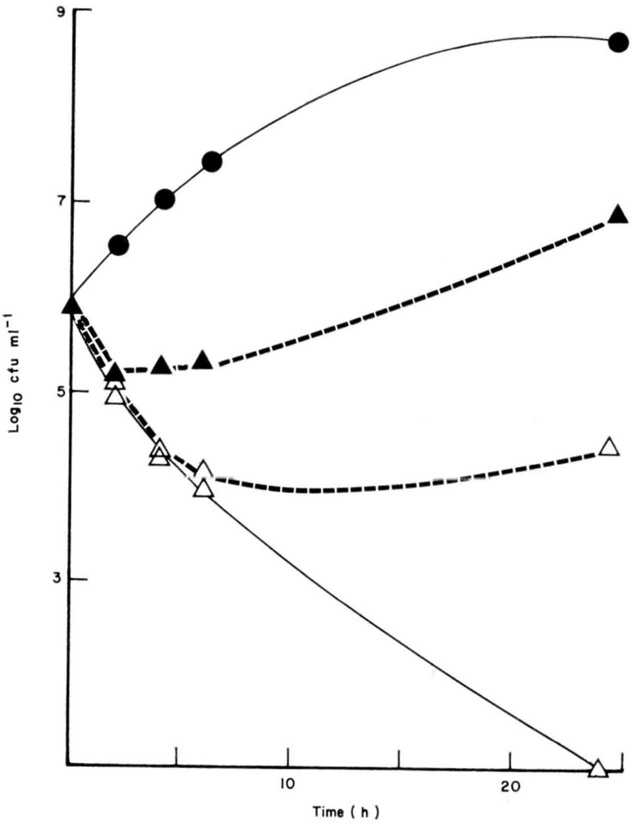

FIG. 3. Effect of different concentrations of catalase on the LP system in glucose-salts medium. Bacterial count (●—●); bacterial count in the presence of SCN^-, 0·17 mM, glucose oxidase, 0·1 u ml^{-1} and LP, 1·5 u ml^{-1} (△—△); as above with addition of catalase, 75 u ml^{-1} (△ - - - △); as above with addition of catalase, 150 u ml^{-1} (▲ - - - ▲).

amino acids is inhibited. Subsequently, protein, DNA and RNA synthesis is inhibited (Table 1) but these effects are secondary to the membrane damage which appears to be reversible for up to 30 min.

Bactericidal and bacteriostatic activity of the LP system against Ps. fluorescens *in milk kept at 4°C*

The above experimental results were obtained in synthetic medium at 30 and 37°C. The experiment illustrated in Fig. 7 shows that *Ps. fluorescens* is also killed by the LP system in milk kept at 4°C. At a fixed rate of H_2O_2

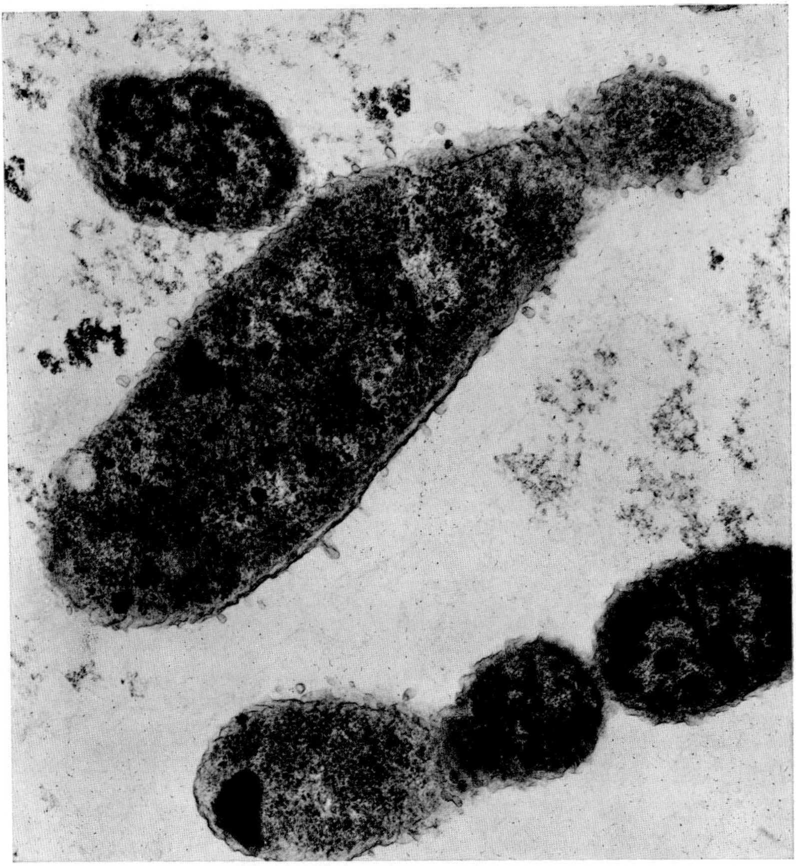

FIG. 4. *Pseudomonas* AR11 after exposure to LP system for 2 h. Cells were collected on a Millipore filter (0·22 μm pore size) covered with a thin layer of agar and fixed for 1 h in 0·2 M cacodylate HCl buffered 3% glutaraldehyde. The filter was washed in buffer, fixed for 1 h in 1% OsO_4 and stained with 1% uranyl acetate for 30 min. After dehydration in a graded series of alcohol-water mixtures it was embedded in Araldite (by courtesy of B. E. Brooker and D. E. Hobbs, NIRD).

generation by glucose oxidase, the degree of killing depends on the level of SCN^-. As expected, at 4°C the rate of killing is slower and the low level of surviving organisms is maintained at the low temperature for a long time, whilst at 30°C the organisms start to multiply after all the SCN^- is oxidized. At 30°C the free H_2O_2 is clearly insufficient to prevent outgrowth of the residual organisms because of the catalase present in milk. The bacteriostasis at 4°C continues for about another 24 h, not because

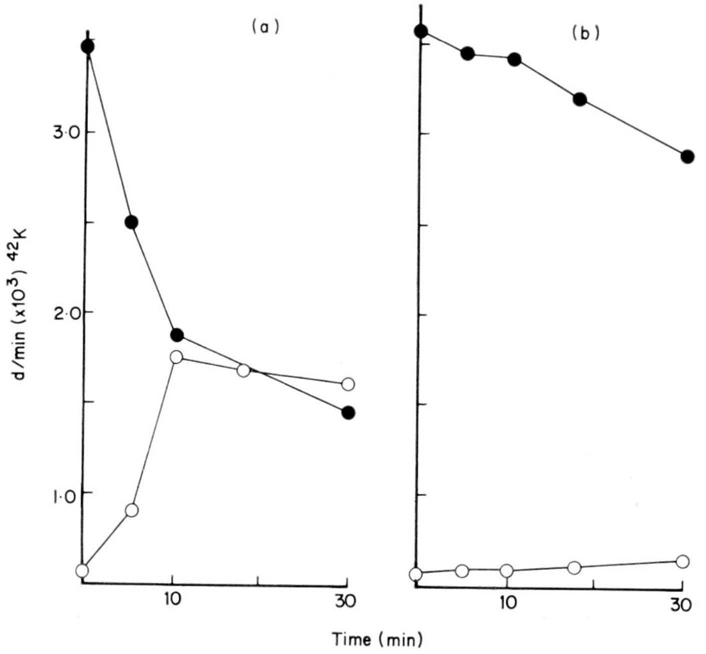

FIG. 5. Leakage of K^+ from *E. coli* (a) in the presence, (b) in the absence of the LP system in glucose-salts medium at 37°C. For details of LP system see Fig. 3. $^{42}K^+$ associated with the organisms after trapping on a Millipore filter (0·45 μm) (●); $^{42}K^+$ present in filtrated glucose-salts medium (○). ^{42}K is expressed in decays per minute (dpm).

the LP system remains active or because of free H_2O_2, but because of the low temperature. Psychrotrophs begin to multiply only after a long lag phase.

The organisms surviving the initial bactericidal activity of the LP system are not resistant. O. Aule (pers. comm.) has shown that subculturing of the residual organisms and a second exposure to the LP system produced the same rate of killing; this remained so even after 16 subculturings.

The Prevention by the LP System of Spoilage of Cheese made from Milk contaminated with *Ps. fluorescens*

This trial was conducted at the experimental dairy of this Institute using milk from our herd. Because the milk hygiene is better than average, the milk was inoculated with a psychrotroph which would multiply

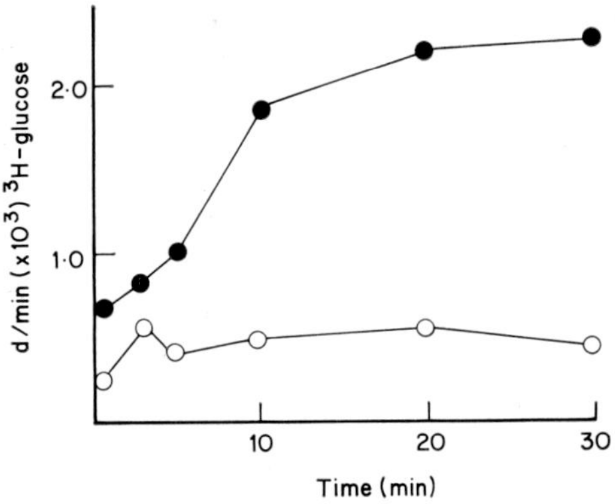

FIG 6. Effect of LP system on uptake of glucose by *E. coli* in glucose-salts medium at 37°C. Glucose uptake in absence of LP system (●); glucose uptake in presence of LP system (○) LP system as in Fig. 3. Uptake was measured using ^{3}H-glucose and radioactivity is expressed in decays per minute.

to at least 10^7 organisms ml^{-1}, thus producing sufficient lipase to cause rancidity in the cheese. The strain used was *Ps. fluorescens* AR11, isolated by Dr. T. Fryer (unpublished) from bulk milk. The lipase production compared well with a previously investigated *Ps. fragi* (Lawrence *et al.*, 1967a) when assayed by the same diffusion method (Lawrence *et al.*, 1967b) and was shown to be heat resistant, approx. 60% of the original lipase activity surviving pasteurization (Law *et al.*, 1976).

Milk was cooled to 5°C and stored in two insulated 50 gallon tanks. Each was inoculated with *Ps. fluorescens* AR11 to give a count above 10^5 ml^{-1}. As expected there was little or no multiplication of the organisms for up to 2 d of storage. On the second day glucose, glucose oxidase and SCN^- were added to the milk of one tank to give 0·1 u ml^{-1} of enzyme, 0·3 mg ml^{-1} of glucose and 0·15 mM SCN^-. As Table 2 shows, at day 3 about 99% of the original inoculum was killed by the LP system while the bacteria in the control multiplied appreciably. At day 4, the count in the treated milk remained static while the control milk contained more than 10^7 organisms ml^{-1}.

At day 4, both milks were pasteurized, the count in the control was reduced to $1{\cdot}2 \times 10^2$ and in the treated milk to < 10 ml^{-1}. The counts were made on crystal violet agar but none of the surviving colonies fluoresced under u.v. light, a characteristic of *Ps. fluorescens*. Before pasteurization, all colonies plated from the control milk fluoresced, but none from the treated milk.

Cheddar cheese was made by the standard method employed at this

TABLE 1. Effect of the LP system on the incorporation of ^{14}C-leucine, ^{14}C-uracil and ^{3}H-thymidine into protein, RNA and DNA respectively

Strain of *E. coli*	Presence[a] or absence of system	Labelled substrate	d min^{-1} after treatment for 30 min	2 h	4 h
9703	—	Leucine	365	1698	4652
9703	+	Leucine	145	278	312
9703	—	Uracil	725	1372	2502
9703	+	Uracil	51	81	209
B/r thy^{-}	—	Thymidine	51	213	1349
B/r thy^{-}	+	Thymidine	11	10	12

[a] 100 μl leucine (60 mCi mmol^{-1}), uracil (61 mCi mmol^{-1}) or thymidine (18·5 Ci mmol^{-1}) was added to 10 ml synthetic medium containing components of the LP system: LP, 1·5 u ml^{-1}; GO, 0·1 u ml^{-1}; glucose, 0·3% (w/v) and SCN, 0·2 mM.

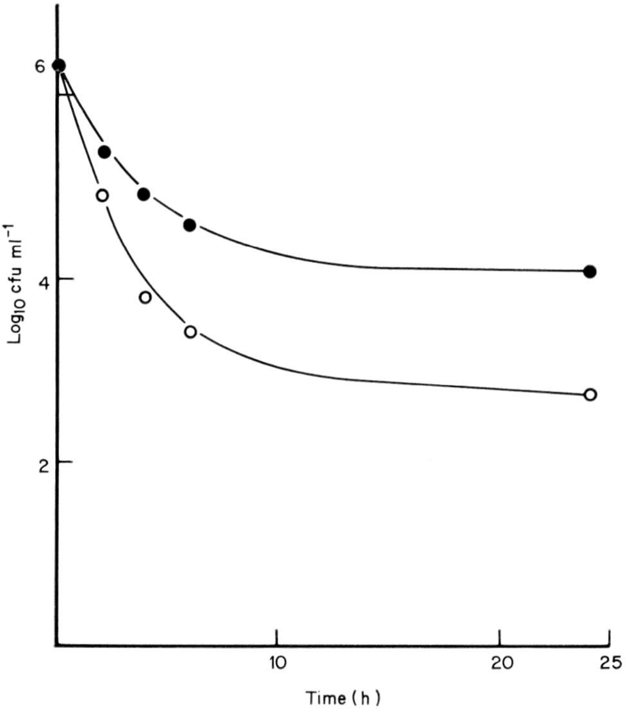

FIG. 7. Effect of the LP system on *Ps. fluorescens* in milk at 4°C. Glucose (0·3%) and glucose oxidase (0·1 u ml^{-1}) were added to aseptically drawn milk. *Ps. fluorescens* in the presence of 0·17 mM SCN^{-} (●); *Ps. fluorescens* in the presence of 0·26 mM SCN^{-} (○).

TABLE 2. Effect of LP system on multiplication of *Ps. fluorescens* and cheese quality

Treatment of milk	No. of *Ps. fluorescens* (cfu $ml^{-1} \times 10^4$) in milk stored at 5°C for (d)				Cheese quality at 4 months	
	0	1	2	3	FFA (μmol 10 g^{-1})	Flavour assessment
None	15	29	150	1400	248	Rancid
LP system	21	1·1	0·2	0·1	50	Normal

Institute and the cheese stored for 4 months. They were then assessed by a tasting panel of 18 trained persons. The cheeses made from the LP treated milk were unanimously declared of average (normal) flavour, but the cheeses made from untreated milk were rancid. This was not surprising because the free fatty acids (FFA) in these cheeses differed widely, nearly 5 times as much FFA 10 g^{-1} of fat occurred in the rancid cheese (Table 2). Low levels of FFA occur in all cheeses after some months of maturation and are known to be derived from the native lipases of milk as FFA have been shown to appear in cheeses made without bacteria from aseptically drawn milk (Reiter *et al.*, 1969).

Similar experiments were also conducted using commercial tanker milk (6 trials) with basically the same results. The inoculum of the NIRD milk with $\sim 10^5$ psychrotrophs ml^{-1} was not excessive because tanker milk regularly contains $2–4 \times 10^5$ Gram negative rods, the majority of which are psychrotrophs (Law *et al.*, 1976).

Conclusions

From this work it appears that it is feasible to supress psychrotrophs by activation of the native LP system. Although it would be impracticable to add glucose and glucose oxidase to milk as a source of H_2O_2 on a commercial scale, the principle has nevertheless been established. Other methods, e.g. coupling the enzyme to a matrix (Björck and Rosén, 1976) have been developed and a patent has been granted for the process (Rosén and Björck, 1975).

Acknowledgements

We wish to thank our colleagues Mrs. E. L. Tzanetaki, Dr. B. A. Law, Dr. M. E. Sharpe and Miss H. R. Chapman for undertaking the cheese-making experiments and their bacteriological control.

References

Aune, T. M. & Thomas, E. L. (1977). Accumulation of hypothiocyanate ion during peroxidase-catalysed oxidation of thiocyanate ion. *European Journal of Biochemistry* **80,** 209–214.

Björck, L. & Rosén, C-G. (1976). An immobilized two-enzyme system for the activation of the lactoperoxidase antibacterial system in milk. *Biotechnology and Bioengineering* **18,** 1463–1472.

Björck, L., Rosén, C-G., Marshall, V. & Reiter, B. (1975). Antibacterial activity of the lactoperoxidase system in milk against pseudomonads and other Gram-negative bacteria. *Applied Microbiology* **30,** 199–204.

BOULANGÉ, M. (1959). Fluctuation du taux des thiocyanates dans le lait frais de vache. *Compte Rendu des Séances de la Société de Biologie*, Paris **153,** 2019–2020.

HOGG, D. M. & JAGO, G. R. (1970). The antibacterial action of lactoperoxidase. The nature of the antibacterial inhibitor. *Biochemical Journal* **117,** 779–790.

HOOGENDORN, H., PIESSENS, J. P., SCHOLTS, W. & STODDARD, C. A. (1977). Hypothiocyanate ion: the inhibitor formed by the system lactoperoxidase-thiocyanate-hydrogen peroxide. *Caries Research* **11,** 77–84.

LAW, B. A., SHARPE, M. E. & CHAPMAN, H. R. (1976). The effect of lipolytic Gram-negative psychrotrophs in stored milk on the development of rancidity in Cheddar cheese. *Journal of Dairy Research* **43,** 459–468.

LAW, B. A., ANDREWS, A. T. & SHARPE, M. E. (1977). Gelation of ultra-high temperature sterilized milk by proteases from a strain of *Pseudomonas fluorescens* isolated from raw milk. *Journal of Dairy Research* **44,** 145–148.

LAWRENCE, R. C., FRYER, T. F. & REITER, B. (1967a). The production and characterization of lipases from a micrococcus and a pseudomonad. *Journal of General Microbiology* **48,** 401–418.

LAWRENCE, R. C., FRYER, T. F. & REITER, B. (1967b). Rapid method for the quantitative estimation of microbial lipases. *Nature, London* **213,** 1264–1265.

MARSHALL, V. M. (1978). *In vitro* and *in vivo* studies on the effect of the lacto-peroxidase/thiocyanate/hydrogen peroxide system on *Escherichia coli*. PhD thesis, University of Reading.

MARSHALL, V. M. & REITER, B. (1976). The effect of the lactoperoxidase/thio-cyanate/hydrogen peroxide system on the metabolism of *Escherichia coli*. *Proceedings of the Society for General Microbiology* **3,** 109.

ORAM, J. D. & REITER, B. (1966). The inhibition of streptococci by lactoperoxidase thiocyanate and hydrogen peroxide. The oxidation of thiocyanate and the nature of the inhibitory compound. *Biochemical Journal* **100,** 382–388.

REITER, B. (1976). Bacterial inhibitors in milk and other biological secretions, with special reference to the complement/antibody, transferrin/lactoferrin and lactoperoxidase/thiocyanate/hydrogen peroxide systems. In *Inhibition and Inactivation of Vegetative Microbes* (Skinner, F. A. & Hugo, W. B., eds). Society for Applied Bacteriology Symposium Series No. 5. London and New York: Academic Press, pp. 31–60.

REITER, B. (1978). Review of the progress of dairy science: antimicrobial systems in milk. *Journal of Dairy Research* **45,** 131–147.

REITER, B. & ORAM, J. D. (1967). Bacterial inhibitors in milk and other biological fluids. *Nature, London* **216,** 328–330.

REITER, B., SOROKIN, Y., PICKERING, A. & HALL, A. J. (1969). Hydrolysis of fat and protein in small cheeses made under aseptic conditions. *Journal of Dairy Research* **36,** 65–76.

REITER, B., MARSHALL, V. M., BJÖRCK, L. & ROSÉN, C-G. (1976). The non-specific bactericidal activity of the lactoperoxidase/thiocyanate/hydrogen peroxide system of milk against *E. coli* and some Gram negative pathogens. *Infection and Immunity* **13,** 800–807.

ROSÉN, C-G. & BJÖRCK, L. (1975). *Swedish patent* **37,** 5224.

Subject Index